Introduction to Electrical Installation Work

Compulsory Units for the 2330 Certificate
in Electrotechnical Technology Level 2
(Installation Route)

Introduction to Electrical Installation Work

Compulsory Units for the 2330 Certificate in Electrotechnical Technology Level 2 (Installation Route)

Trevor Linsley

ELSEVIER

AMSTERDAM · BOSTON · HEIDELBERG · LONDON · NEW YORK · OXFORD
PARIS · SAN DIEGO · SAN FRANCISCO · SINGAPORE · SYDNEY · TOKYO

Newnes is an imprint of Elsevier

Newnes

Newnes is an imprint of Elsevier
Linacre House, Jordan Hill, Oxford OX2 8DP, UK
30 Corporate Drive, Suite 400, Burlington MA 01803, USA

First Edition 2007

British Library Cataloguing in Publication Data
A catalogue record for this book is available from the British Library

Library of Congress Cataloguing in Publication Data
A catalogue record for this book is available from the Library of Congress

ISBN 13: 978-0-75-068114-8
ISBN 10: 0-75-068114-4

For information on all Newnes publications visit our website at www.books.elsevier.com

Typeset by Charon Tec Ltd (A Macmillan Company), Chennai, India
www.charontec.com

Printed and bound in Italy

07 08 09 10 11   10 9 8 7 6 5 4 3 2 1

Contents

Preface

Introduction to Electrical Installation Work is, as the title implies, a first book of electrical installation practice. It is designed to be a simple introduction to electrical theory and practice and, therefore, does not contain any difficult mathematics or complicated electrical theory.

The book will be of assistance to students taking a first year electrical course, particularly those taking the City & Guilds 2330 Level 2 Certificate in Electrotechnial Technology.

Introduction to Electrical Installation Work provides a sound basic knowledge of electrical practice which will also be valuable to the other trades in the construction industry, particularly those involved in multi skilling activities.

This book incorporates the requirements of the latest regulations, particularly:

◆ The IEE Wiring Regulations (BS 7671:2001) incorporating Amendments 1:2002 and 2:2004.
◆ The New Work at Height Regulations 2005
◆ The New Part P Building Regulations (Electrical Safety in Dwellings) 2005
◆ The New (Harmonised) Fixed Cable Core Colours (2006)

The City & Guilds course is in four units which correspond to the four chapters in this book. Each chapter concludes with Assessment Questions in preparation for the City & Guilds On-Line Assessment.

I would like to acknowledge the assistance given by the following manufacturers and organisations in the preparation of this book:

Crabtree Electrical Industries Ltd
R.S. Components Ltd
The Institution of Electrical Engineers
The British Standards Institution
The City & Guilds of London Institute
Stocksigns Ltd
Wylex Electrical Components

I would also like to thank my colleagues at Blackpool and The Fylde College and the proposal reviewers for their suggestions and advice during the preparation of this book.

Finally, I would like to thank Joyce, Samantha and Victoria for their support and encouragement.

Trevor Linsley

CHAPTER 1

Working Effectively and Safely in an Electrical Environment

Chapter 1 covers the topics described in the first core unit of the City & Guilds 2330 Syllabus for the Level 2 Certificate in Electrotechnical Technology

This Chapter describes the requirements that are essential to enable electrotechnical activities to be carried out safely and effectively within the parameters set by the current safety legislation and best practice related to the Electrotechnical Industry.

Laws and Safety Regulations

The construction industry is one of the biggest industries in the United Kingdom, although most workers are employed by small companies employing less than 25 people.

The construction industry carries out all types of building work from basic housing to offices, hotels, schools and airports.

In all of these construction projects the Electrotechnical Industry plays a major role in designing and installing the electrical systems to meet the needs of those who will use the completed buildings.

The construction process is potentially hazardous and many construction sites these days insist on basic safety standards being met before you are allowed on site. All workers must wear hard hats and safety boots or safety trainers and use low voltage or battery tools. When the building project is finished, all safety systems will be in place and the building will be safe for those who will use it. However, during the construction period, temporary safety systems are in place. People work from scaffold towers, ladders and stepladders. Permanent stairways and safety handrails must be put in by the construction workers themselves.

When the electrical team arrives on site to, let us say, 'first fix' a new domestic dwelling house, the downstairs floorboards and the ceiling plasterboards will probably not be in place, and the person putting in the power cables for the downstairs sockets will need to step over the floor joists, or walk and kneel on planks temporarily laid over the floor joists.

The electrical team spend a lot of time on their hands and knees in confined spaces, on ladders, scaffold towers and on temporary safety systems during the 'first fix' of the process and, as a consequence, slips, trips and falls do occur.

To make all working environments safer, laws and safety regulations have been introduced. To make **your**

There are lots of safety regulations in this job

working environment safe for yourself and those around you, you must obey all the safety regulations that are relevant to your work.

The many laws and regulations controlling the working environment have one common purpose, to make the working environment safe for everyone.

Let us now look at some of these laws and regulations as they apply to the Electrotechnical Industry.

Statutory Laws

Acts of Parliament are made up of Statutes. Statutory Laws and Regulations have been passed by Parliament and have therefore become laws. The City and Guilds Syllabus requires that we look at seven Statutory Regulations.

1. The Health & Safety at Work Act 1974

There are seven Statutory Laws

◆ The purpose of the HSAWA is to provide the legal framework for stimulating and encouraging high standards of health and safety at work.

◆ The Act places the responsibility for safety at work on **both** workers and employers.

◆ The HSAWA is an "Enabling Act" which allows the Secretary of State to make further regulations and modify existing regulations to create a safe working environment without the need to pass another Act of Parliament.

2. Electricity at Work Regulations 1989

◆ These Regulations are made under the Health & Safety at Work Act and are enforced by the Health & Safety Executive (HSE).

◆ The purpose of the Regulations is to "require precautions to be taken against the risk of death or personal injury from electricity in work activities".

◆ An electrical installation wired in accordance with the IEE Regulations BS 7671 will also meet the requirements of the EWR.

3. The Electricity Safety, Quality and Continuity Regulations 2002

◆ These Regulations are designed to ensure a proper and safe supply of electrical energy up to the consumer's mains electrical intake position.

◆ They will not normally concern the electrical contractor, except in that it is these Regulations which set out the earthing requirements of the supply.

4. The Management of Health & Safety at Work Regulations 1999

◆ To comply with the Health & Safety at Work Act 1974 employers must have "robust health and safety systems and procedures in the workplace".

◆ Employers must "systematically examine the workplace, the work activity and the management of safety through a process of risk assessment".

◆ Information based upon the risk assessment findings must be communicated to relevant staff.

◆ So, risk assessment must form a part of any employer's "robust policy of health and safety".

5. Provision and Use of Work Equipment Regulations 1998

◆ These Regulations place a general duty of care upon employers to ensure minimum requirements of plant and equipment used in work activities.

◆ If an employer has purchased good quality plant and equipment, and that plant and equipment is well maintained, there is little else to do.

6. COSHH Regulations (2002)

◆ The Control of Substances Hazardous to Health Regulations (COSHH) control people's exposure to hazardous substances in the workplace.

◆ Employers must carry out risk assessments and, where necessary, provide PPE (Personal Protective Equipment) so that employees will not endanger themselves.

◆ Employees must also receive information and training in the safe storage, disposal and emergency procedures which are to be followed by anyone using hazardous substances.

7. Personal Protective Equipment Regulations (PPE)

◆ PPE is defined as all equipment designed to be worn or held in order to protect against a risk to health and safety.

◆ This includes most types of protective clothing and equipment such as eye, foot and head protection, safety harnesses, life jackets and high visibility clothing.

◆ Employers must provide PPE free of charge and employees must make use of it for their protection.

◆ Figure 1.1 below shows the type of safety signs which might be used to indicate the type of PPE to be worn in particular circumstances for your protection.

(a)   (b)

(c)   (d)

(e)   (f)

(g)   (h)

Figure 1.1 Safety Signs showing type of PPE to be worn

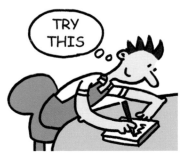

TRY THIS

Have you seen these or any other PPE signs at work? Make a list of the PPE signs that you have seen at work and state why they were important in that particular work situation

Non-Statutory Laws are still very important

Non-Statutory Regulations

Statute Law is law which has been laid down by Parliament as Act of Parliament.

Non-Statutory Regulations and codes of practice interpret the Statutory Regulations.

Non-Statutory **does not** mean non-compulsory. If the Non-Statutory Regulation is relevant to your part of the Electrotechnical Industry then you **must** comply.

The City & Guild Syllabus requires us to look at only one Non-Statutory Regulation, the IEE Regulations.

The IEE Wiring Regulations, the requirements for Electrical Installations (BS 7671)

◆ The IEE Wiring Regulations relate principally to the design, selection, erection, inspection and testing of electrical installations. The Regulations are Non-Statutory Regulations but are recognised as the National Standards for electrical installation work in the United Kingdom.

◆ **They apply to:**
  ▌ permanent or temporary installations
  ▌ in and about buildings generally
  ▌ to agricultural and horticultural premises
  ▌ to construction sites
  ▌ and to caravans and caravan sites

◆ They are the "Electricians' Bible" and provide authoritative framework for all work activities undertaken by electricians.

◆ If your work meets the requirements of the IEE Regulations, it will also comply with the Statutory Regulations

Health and Safety Responsibilities

Everyone has a duty of care under the Health and Safety at Work Laws and Regulations to take care of themselves and others who may be affected by their work activities.

In general terms the employer must put adequate health and safety systems in place at work and an employee (worker) must use all safety systems and procedures responsibly. In more specific terms, there are **twelve** actions that an employer must take to comply with the Health and Safety Laws.

Your health and safety at work is **very** important

An Employer must:

◆ make the workplace safe and without risk to health

◆ provide a Health & Safety Policy Statement if there are more than **five** employees

◆ provide adequate information, instruction, training and supervision necessary for the health and safety of all employees

◆ provide any protective clothing or equipment (PPE) required by the Health & Safety Act

◆ report certain injuries, diseases and dangerous occurrences to the enforcing authorities

◆ provide adequate first aid facilities

◆ provide adequate welfare facilities

◆ undertake precautions against fire, provide adequate means of escape and the means of fighting a fire

◆ ensure that plant, equipment and machinery are safe and that safe systems and procedures of work are put in place and followed

◆ ensure that articles and substances are moved, stored and used safely

◆ keep dust, fumes and noise under control

◆ display a current Certificate as required by the Employers' Liability (Compulsory Insurance) Act 1969.

An Employee (Worker) must:

◆ take reasonable care of their own health and safety and that of others who might be affected by what you do, or may not do, while at work

◆ co-operate with your employer on all matters relating to health and safety issues

◆ not interfere with, or mis-use anything provided for health and safety or welfare in the workplace

◆ report any identified health and safety problem in the workplace to a supervisor, manager or employer

Safety Signs

Safety signs are displayed in the working environment to inform workers of the rules and regulations especially relevant to a particular section of the workplace.

They inform and give warning of possible danger and **must be obeyed**.

There are **four types** of safety signs:

1. Warning signs

2. Advisory signs

3. Mandatory signs

4. Prohibition signs

Warning Signs (these give safety information)

These are triangular yellow signs with a black border and symbol as shown in Fig. 1.2.

Figure 1.2   Warning Signs

Advisory Signs (these also give safety information)

Advisory or safe condition signs are square or rectangular green signs with a white symbol as shown in Fig. 1.3. They give information about safety provision.

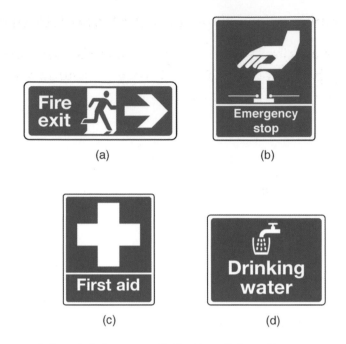

Figure 1.3 Advisory or Safe Condition Signs

Mandatory Signs (these are 'MUST DO' signs)

These are circular blue signs with a white symbol as shown in Fig. 1.4. They give instructions which **must be obeyed**.

Figure 1.4 Mandatory Signs

Prohibition Signs (these are 'MUST NOT DO' signs)

These are circular white signs with a red border and red cross bar as shown in Fig. 1.5. They indicate an activity which **must not** be carried out.

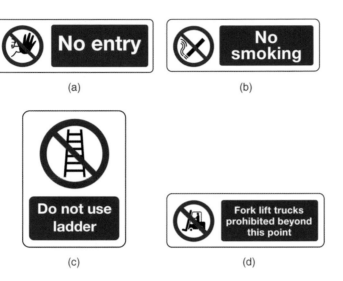

(a)   (b)

(c)   (d)

Figure 1.5   Prohibition Signs

Have you ever seen any of the signs on these two pages when you have been working? Make a list of where you saw them and why they were important in that work situation

Accident and Emergency Procedures

Despite new legislation, improved information, education and training, accidents at work do still happen.

◆ An accident may be defined as any uncontrolled event causing injury or damage to an individual or property.

◆ Make sure that even small accidents at work are recorded in the First Aid/Accident Report book such as that shown in Fig. 1.6.

To avoid having an accident you should:

◆ recognise situations which could lead to an accident and avoid them

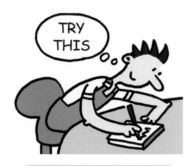

Have you had an accident at work?

Did it hurt? Did you report it? Was it written up in the First Aid/Accident Report book? Was the book data protection compliant with removable sheets?

Figure 1.6   First Aid Logbook/Accident book with data protection compliant removable sheets

◆ follow your Company's safety procedures – for example, fit safety signs when isolating electricity supplies and screen off work areas from the general public

◆ do not misuse or interfere with equipment provided to protect health and safety

◆ dress appropriately and use PPE when necessary

◆ behave appropriately and with care

◆ stay alert and avoid fatigue

◆ always work within your level of competence

◆ take a positive decision to act and work safely. A simple accident may prevent you from working or following your favourite sport or hobby

Emergency Procedures – Fire Control

Fires in industry damage property and materials, injure people and sometimes cause loss of life. Everyone should make an effort to prevent fires, but those which do break out should be extinguished as quickly as possible.

In the event of a fire you should:

◆ raise the alarm

◆ turn off machinery, gas and electricity supplies in the area of the fire

◆ close doors and windows but without locking or bolting them

◆ remove combustible material away from the path of the fire if this can be done safely

◆ attack small fires with the correct extinguisher

> *Only attack the fire if you can do so without endangering your own safety in any way*

Fires are divided into Four Classes or Categories

◆ Class A are wood, paper and textile fires

◆ Class B are liquid fires such as paint, petrol and oil

◆ Class C are fires involving gas or spilled liquefied gas

◆ Class D are very special types of fire involving burning metal

Electrical fires do not have a special category because, once started, they can be identified as one of the four above types.

Fire extinguishers are for dealing with small fires and different types of fire must be attacked with a different type of extinguisher.

Figure 1.7 shows the correct type of extinguisher to be used on the various categories of fire. The colour coding shown is in accordance with BS EN3:1996.

Type of fire extinguisher / Type of fire	(i) Water — Signal red flash on red	(ii) Foam — Pale cream flash on red	(iii) Carbon dioxide gas — Black flash on red	(iv) Dry powder — French blue flash on red	(v) Vapourizing foam — Emerald green flash on red
CLASS A. Paper, Wood and Fabric	✓ Yes	✓ Yes	✗ No	✓ Yes	✓ Yes
CLASS B. Flammable Liquids	✗ No	✓ Yes	✓ Yes	✓ Yes	✓ Yes
CLASS C. Flammable Gases	✗ No	✗ No	✓ Yes	✓ Yes	✓ Yes
Electrical fires	✗ No	✗ No	✓ Yes	✓ Yes	✓ Yes
Motor vehicle protection	✗ No	✓ Yes	✗ No	✓ Yes	✓ Yes

Figure 1.7   Fire Extinguishers and their Applications

Emergency Procedures – Electric Shock

◆ Electric shock occurs when a person becomes part of the electrical circuit as shown in Fig. 1.8.

Figure 1.8   Touching a Live Conductor can make a Person part of the Electrical Circuit and may lead to an Electric Shock

◆ The level or intensity of the shock will depend upon many factors such as age, fitness and the circumstances in which the shock is received.

◆ The lethal level is approximately 50 mA, above which muscles contract, the heart flutters and breathing becomes difficult.

◆ Below 50 mA only an unpleasant tingling sensation may be experienced or you may feel like you have been struck very hard in the chest.

> *To prevent people receiving an electric shock accidentally, all circuits must contain protective devices and all exposed metal must be earthed*

> *All circuits must be electrically isolated before any work is carried out*

Actions to be taken upon finding a Workmate receiving an Electric Shock are as follows:

◆ Switch off the supply if possible

◆ Alternatively, remove person from the supply *without touching him*

◆ If breathing or heart has stopped, immediately call professional help by dialling the Emergency Number 999 (or 112) and asking for the ambulance service. Give precise directions to the scene of the accident. The casualty stands the best chance of survival if the Emergency Services can get a rapid response Paramedic Team quickly to the scene. They have extensive training and will have specialist equipment with them

◆ Only then should you apply resuscitation or cardiac massage until the patient recovers or help arrives.

Emergency Procedures – First Aid

Despite all the Health and Safety Laws and Regulations and despite all the safety precautions taken in the working environment to prevent injury to the workforce, accidents do happen at work and if a workmate is injured you will want to help. If you are not a qualified First Aider, limit your actions to the obvious common sense assistance and get help from someone who is qualified.

Figure 1.9   Accidents do happen at Work

Let us now look at some First Aid procedures which should be practised under expert guidance before they are required in an emergency.

Bleeding

If the wound is dirty, rinse it under clean running water. Clean the skin around the wound and apply a plaster, pulling the skin together.

If the bleeding is severe, apply direct pressure to reduce the bleeding. If the injury is to a limb then raise the limb if possible. Apply a sterile dressing or pad and bandage firmly before obtaining professional advice.

To avoid possible contact with hepatitis or the AIDS virus, when dealing with open wounds, First Aiders should avoid contact with fresh blood by wearing plastic or rubber protective gloves, or by allowing the casualty to apply pressure to the bleeding wound.

Burns

Remove heat from the burn to relieve the pain by placing the injured part under clean cold water if at all possible. Do not remove burnt clothing sticking to the skin. Do not apply lotions or ointments. Do not break blisters or attempt to remove loose skin. Cover the injured area with a clean dry dressing.

Broken Bones

Make the casualty as comfortable as possible by supporting a broken limb, either by hand or with padding. **Do not move** the casualty unless by remaining in that position they are likely to suffer further injury. Obtain professional help as soon as possible.

Contact with Chemicals

Wash the affected area very thoroughly with clean cold water. Remove any contaminated clothing. Cover the affected area with a clean sterile dressing and seek expert advice. It is a wise precaution to treat all chemical substances as possibly harmful; even commonly used substances can be dangerous if contamination is from concentrated solutions. When handling dangerous substances it is also good practice to have a neutralising agent to hand.

Exposure to Toxic Fumes

Get the casualty into fresh air quickly and encourage deep breathing if conscious. Resuscitate if breathing has stopped. Obtain expert medical advice as fumes may cause irritation of the lungs.

Sprains and Bruising

A cold compress can help to relieve swelling and pain. Soak a towel or cloth in cold water, squeeze it out and place it on the injured part. Renew the compress every few minutes.

Breathing stopped

Remove any restrictions from the face and any vomit, loose or false teeth from the mouth. Loosen tight clothing around the neck, chest and waist. To ensure a good airway, lay the casualty on his back and support the shoulders on some padding. Tilt the head backwards and open the mouth. If the casualty is faintly breathing, lifting the tongue clear of the airway may be all that is necessary to restore normal breathing. However, if the casualty does not begin to breathe, open your mouth wide and take a deep breath, close the casualty's nose by pinching with your fingers and, sealing your lips around his mouth, blow into his lungs until the chest rises. Remove your mouth and watch the casualty's chest fall. Continue this procedure at your natural breathing rate. If the mouth is damaged or you have difficulty making a seal around the casualty's mouth, close his mouth and inflate the lungs through the nostrils. Give artificial respiration until natural breathing is restored or until professional help arrives.

Heart stopped beating

This sometimes happens following a severe electric shock. If the casualty's lips are blue, the pupils of his eyes widely dilated and the pulse in the neck cannot be felt then he may have gone into cardiac arrest. Act quickly and lay the casualty on his back. Kneel down beside him and place the heel of your hand in the centre of his chest. Cover this hand with your other hand and interlace the fingers. Straighten your arms and press down on his chest sharply with the heel of your hands and then release the pressure. Continue to do this **15** times at the rate of **one push per second**. Check the casualty's pulse. If none is felt, give **two breaths** of artificial respiration and then a further **15** chest compressions. Continue this procedure until the heartbeat is restored and the artificial respiration until normal breathing returns. Pay close attention to the condition of the casualty while giving the heart massage. When a pulse is restored the blueness around the mouth will quickly go away and you should stop the heart massage. Look carefully at the rate of breathing. When this is also normal, stop giving artificial respiration. Treat the casualty for shock, place him in the recovery position and obtain professional help.

Shock

Everyone suffers from shock following an accident. The severity of the shock depends upon the nature and extent of the injury. In cases of severe shock the casualty will become pale and his skin become clammy from sweating. He may feel faint, have blurred vision, feel sick and complain of thirst. Reassure the casualty that everything that needs to be done is being done. Loosen tight clothing and keep the casualty warm and dry until help arrives. **Do not** move the casualty unnecessarily or give anything to drink.

Finally, remember that every accident must be reported to an employer and the details of the accident and treatment given suitably documented. A First Aid Logbook or Accident Report book such as that shown in Fig. 1.6 above containing first aid treatment record sheets could be used to effectively document such accidents that occur in the workplace and the treatment given. Failure to do so may influence the payment of compensation at a later date if an injury leads to permanent disability. To comply with the Data Protection Regulations, from the 31st December 2003 all First Aid Treatment Logbooks or Accident Report books must contain perforated sheets which can be removed after completion and filed away for personal security.

Emergency Procedures – Electrical Isolation and Lock Off

◆ The IEE Regulations tell us that every circuit must be provided with a means of isolation.

◆ The Electricity at Work Regulations tell us that before work commences on electrical equipment it must be disconnected from the source of supply and that the disconnection must be secure.

◆ A small padlock will ensure the security of the disconnection, or the fuse or MCB may be

Electrical isolation is an important safety procedure – practice this technique at work under the guidance of your Supervisor removed and kept in a safe place whilst work is carried out.

◆   Where a test instrument or voltage indicator is used to prove the supply dead, the same device must be tested to prove it is still working. Figure 1.10 shows a typical voltage indicator and Fig. 1.11 shows a typical voltage proving unit.

◆   The test leads and probes of the test instrument must comply with the Health & Safety Executive Guidance Note 38, giving adequate protection to the user. These are robust leads with finger shields.

Figure 1.10   Typical voltage indicator

Figure 1.11   Voltage proving unit

◆   To deter anyone from reconnecting the supply,
     a notice must be fixed on the isolator saying
     'Danger – Electrician at Work'.

A suitable electrical isolation procedure is shown in
Fig. 1.12, which you should practice in the workshop
under the guidance of your Lecturer or at work under
the guidance of your Supervisor. Electrical isolation is
an important safety procedure.

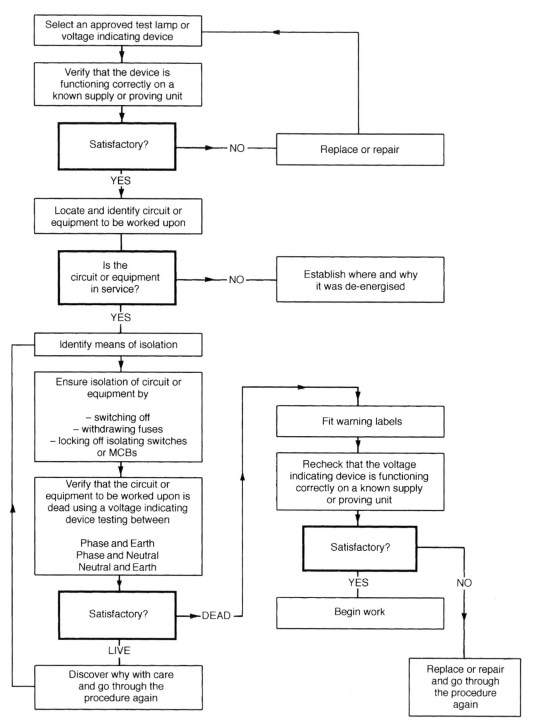

Figure 1.12   Flowchart for a secure isolation procedure

Organisations having Electrotechnical Activities

When we talk about the Electrotechnical Industry we are referring to all those different organisations or companies which provide an electrical service of some kind.

Electrical contractors install equipment and systems in new buildings. Once a building is fully operational the electrical contractor may provide a maintenance service to that client or customer or alternatively the client may employ an 'in-house' electrician to maintain the installed electrical equipment. It all depends on the amount of work to be done and the complexity of the customer's systems.

The City & Guilds Syllabus directs us to look at **twelve** different organisations having electrotechnical activities and ten services provided by the electrotechnical industry, so here goes.

1. Electrical Contractors

◆ Electrical contractors provide a design and installation service for all types of buildings and construction projects

◆ The focus of this type of organisation is on all types of electrotechnical activities in and around buildings

◆ They install electrical equipment

◆ They install electrical wiring systems

◆ They carry out their installation work in domestic, commercial, industrial, agricultural and horticultural buildings

The Electrotechnical Industry is made up of many different organisations or companies

2. Factories

◆ Factories contain lots of electrical plant and equipment

◆ The wheels of all types of industry are driven by electromechanical devices and electrotechnical activities

3. Process Plants

◆ Whether they process food or nuclear fuels, the prime mover for all processes is electrical plant, control and instrumentation equipment and machine drives

4. Local Councils

◆ Local Councils are responsible for many different types of community buildings from Town Halls to Swimming Pools

◆ The buildings all have electrical systems which require installation, maintenance and repair

> **TRY THIS**
>
> Which type of Electrotechnical organisation do you belong to?
>
> What types of work have you carried out so far?

5. Commercial Buildings and Complexes

◆ The 'office type' activities carried out in these buildings require that electrical communication and data transmission systems are installed, maintained and repaired

6. Leisure Centres

◆ These type of buildings contain lots of equipment driven by human sweat but which is also controlled and monitored by electrical and electronic systems

◆ Leisure centres might contain a swimming pool or 'hot-air' sauna. Both types of electrical installation are considered 'Special Installations' by the IEE Regulations BS:7671

7. Panel Builders

◆ Panel Builders build specialist control, protection and isolation main switchgear systems for commerce and industry

◆ The panel incorporates the isolation and protection systems required by the electrical installation

8. Motor Re-wind and Repair

◆ Electrical motors and their drives usually form an integral part of the industrial system or process

◆ Electrical motors and transformers sometimes break down or burn out

◆ An exact new replacement can often be quickly installed

◆ Alternatively, the existing motor can be re-wound and reconditioned by a specialist company if time permits

9. Railways

◆ The prime mover for a modern inter-city type electric train is an electric motor

◆ Electric trains require an infrastructure of electrical transmission lines throughout the network

◆ All rail movements require signal and control systems

◆ Railway station buildings contain electrical and electronic installations

10. The Armed Forces

◆ The Armed Forces operate in harsh, hostile and unpredictable environments

◆ They need to adapt, modify and repair electrical and electronic systems in a war situation away from their home base and a comfortable well equipped workshop

◆ A modern warship can contain as many people as an English village. They need electrotechnical systems to support them and to keep them safe twenty-four hours per day, seven days per week

11. Hospitals

◆ Hospitals contain a great deal of high technology equipment

◆ This equipment requires power and electronic systems

◆ Life monitoring equipment must continue to operate in a power failure

◆ Standby electrical supplies are, therefore, often an important part of a hospital's electrical installations

12. Equipment and Machine Manufacturers

◆ White goods, brown goods, computer hardware, motors and transformers are manufactured to meet the increasing demands of the domestic, commercial and industrial markets

◆ They are manufactured to very high standards and often contain very sophisticated electrical and electronic circuits and systems

◆ They manufacture to British and European Standards

Services provided by the Electrotechnical Industry

1. Lighting and power installations

◆ Ensure that the building in which they are installed:
  ▷ is illuminated to an appropriate level
  ▷ is heated to a comfortable level
  ▷ has the power circuits to drive the electrical and electronic equipment required by those who will use the buildings

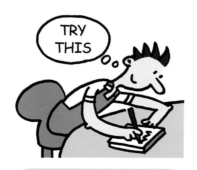

2. Emergency lighting and security systems

◆ These ensure that the building is safe to use in unforeseen or adverse situations

◆ And is secure from unwanted intruders

Which services does the company you work for provide within the Electrotechnical Industry?

3. Building management and control systems

◆ These systems provide a controlled environment for the people who use commercial buildings

◆ They provide a pleasant environment so that people can work effectively and efficiently

4. Instrumentation

◆ Electrical instrumentation allows us to monitor industrial processes and systems often at a safe distance

5. Electrical maintenance

◆ A programme of planned maintenance allows us to maintain the efficiency of all installed systems

6. Live cable jointing

◆ Making connections to 'live' cables provides a means of connecting new installations and services to existing live supply cables without inconvenience to existing supplies caused by electrical shutdown. **This work requires special training**.

7. Highway electrical systems

◆ Illuminated motorways, roads and traffic control systems make our roads and pavements safe for vehicles and pedestrians

8. Electrical panel building

◆ Main electrical panels provide a means of electrical isolation and protection

◆ They also provide a means of monitoring and measuring electrical systems in our commercial and industrial buildings

9. Electrical machine drive installations

◆ Electrical machine drives drive everything that makes our modern life comfortable from
  ▶ trains and trams to
  ▶ lifts and air conditioning units
  ▶ refrigerators, freezers and all types of domestic appliances

10. Consumer and commercial electronics

◆ These give us data processing and number crunching

◆ Electronic mail and access to information on the world wide web

◆ Access to high quality audio and video systems

Roles and Responsibilities of Workers in the Electrotechnical Industry

Any electrotechnical organisation is made up of a group of individuals with various duties, all working together for their own good, the good of their employer and their customers.

There is often no clear distinction between the duties of the individual employees, each do some of the others' work activities.

Responsibilities vary, even by people holding the same job title and some individuals hold more than one job title. However, let us look at some of the roles and responsibilities of those working in the electro-technical industry.

Design Engineer

◆ Will normally meet with clients and other trade professionals to interpret the customers requirements

◆ He or she will produce the design specification which enables the cost of the project to be estimated

Estimator/Cost Engineer

◆ Measures the quantities of labour and material necessary to complete the electrical project using the plans and specifications for the project

◆ From these calculations and the company's fixed costs, a project cost can be agreed

Contracts Manager

◆ May oversee a number of electrical contracts on different sites

◆ Will monitor progress in consultation with the project manager on behalf of the electrical companies

◆ Will cost out variations to the initial contract

◆ May have Health & Safety responsibilities because he or she has an overview of all company employees and contracts in progress

Project Manager

◆ Is responsible for the day to day management of one specific contract

TRY THIS

Where do you fit into your Company's workforce?

What is your job title – Apprentice/ Trainee?

What is your Supervisor's job title?

What is the name of the Foreman?

◆ Will have overall responsibility on that site for the whole electrical installation

◆ Attends site meetings with other trades as the representative of the electrical contractor

Service Manager

◆ Monitors the quality of the service delivered under the terms of the contractor

◆ Checks that the contract targets are being met

◆ Checks that the customer is satisfied with all aspects of the project

◆ The Service Manager's focus is customer specific while the Project Manager's focus is job specific

Technician

◆ Will be more office based than site based

◆ Will carry out surveys of electrical systems

◆ Updates electrical drawings

◆ Obtains quotations from suppliers

◆ Maintains records such as ISO 9000 quality systems

◆ Carries out testing inspections and commissioning of electrical installations

◆ Trouble shoots

Supervisor/Foreman

◆ Will probably be a mature electrician

◆ Has responsibility for small contracts

◆ Has responsibility for a small part of a large contract

◆ Will be the leader of a small team (e.g. electrician and trainee) installing electrical systems

"I am an operative, but hope one day to be the supervisor or foreman"

Operative

◆ Carries out the electrical work under the direction and guidance of a supervisor

◆ Should demonstrate a high degree of skill and competence in electrical work

◆ Will have, or be working towards, a recognised electrical qualification and status as an electrician, approved electrician or electrical technician

Mechanic/Fitter

◆ An operative who usually has a 'core skill' or 'basic skill' and qualification in mechanical rather than electrical engineering

◆ In production or process work he or she would have responsibility for the engineering and fitting aspects of the contract, while the electrician and instrumentation technician would take care of the electrical and instrumentation aspects

◆ All three operatives must work closely in production and process work

◆ 'Additional skilling' or 'multi-skilling' training produces a more flexible operative for production and process plant operations

Maintenance Manager/Engineer

◆ Is responsible for keeping the installed electrotechnical plant and equipment working efficiently

◆ Takes over from the builders and contractors the responsibility of maintaining all plant equipment and systems under his or her control

◆ Might be responsible for a hospital or a commercial building, a university or college complex

◆ Will set up routine and preventative maintenance programmes to reduce possible future breakdowns

◆ When faults or breakdowns do occur he or she will be responsible for the repair using the company's maintenance staff

Professional Bodies supporting Electrotechnical Organisations

If you are reading this book I would guess that you are an electrical trainee working in one sector of the Electrotechnical Industry. You hope to eventually pass the City & Guilds 2330 Parts 2 and 3 qualifications, take your AM2 Practical Assessment and become a qualified electrician. Believe me, I do wish you well, because you are the future of the Electrotechnical Industry.

As a trainee, you are probably employed by an electrical company and attend your local College on either a 'Day Release' or 'Block Release' scheme. The combination of work and College will provide you with the skills you will need to become 'fully qualified'!

So, although you are doing all the work yourself, you are being sponsored or supported by the company that you work for, the JTL (JIB Training Limited) and the City & Guilds of London Institute to become professionally qualified as an electrician.

It is in this same way that the Professional Bodies support the Electrotechnical Industry. They provide a structure of help, support and guidance to the individual companies that make up the Electrotechnical Industry.

So let us look at some of the Professional Bodies which support the Electrotechnical organisations like the company you work for.

The IET (The Institute of Engineering and Technology)

◆ The IET was formed in spring 2006 by bringing together the IEE (Institution of Electrical Engineers) and the IIE (Institution of Incorporated Engineers)

◆ The IET is Europe's largest professional society for engineers

◆ The IET publishes the IEE Wiring Regulations to BS:7671

◆ They also produce many other publications and provide training courses to help electricians, managers and supervisors to keep up to date with the changes in the relevant regulations

◆ The IEE On Site Guide describes the "requirements for electrical installations"

◆ Seven guidance notebooks are available

◆ The Electricians Guide to the Building Regulations clarifies the requirements for electrical operatives of the new Part P Regulations which came into effect on the 1st January 2005

◆ Wiring Matters is a quarterly magazine published by the IET covering many of the topics which may trouble some of us in the Electrotechnical Industries

◆ All of these publications can be purchased by visiting the IET website at www.theiet.org.uk/shop

The ECA (Electrical Contractors Association)

◆ The ECA was founded over 100 years ago and is a Trade Association representing electrotechnical companies

◆ Membership is made up of electrical contracting companies both large and small

◆ Customers employing an electrical contractor who has ECA membership are guaranteed that the work undertaken will meet all relevant regulations. If the work undertaken fails to meet the relevant standards, the ECA will arrange for the work to be rectified at no cost to the customer

◆ The work of the ECA member is regularly assessed by the Association's UKAS accredited inspection body

◆ Those electrotechnical companies which are Members of the ECA are permitted to display the ECA logo on their company vehicles and stationery

◆ Further information can be found on the ECA website at www.eca.co.uk

The National Inspection Council for Electrical Installation Contracting (NICEIC)

◆ The NICEIC is an independent consumer safety organisation, set up to protect users of electricity against the hazards of unsafe electrical installations

◆ It is the electrical industry's safety regulatory body

◆ The NICEIC publishes a list of Approved Contractors whose standard of work is regularly assessed by local area engineers

◆ Customers employing an electrical contractor who has NICEIC membership can be assured that the work carried out will meet all relevant standards. If the work undertaken fails to meet all relevant standards, the name of the electrical contractor will be removed from the "NICEIC Approved List"

◆ Some work, such as local authority work, is only available to NICEIC Approved Contractors

◆ Further information can be found at www.niceic.org.uk

Trade Unions

◆ Trade Unions have a long history of representing workers in industry and commerce

◆ The relevant Unions negotiate with employer organisations the pay and working conditions of their members

◆ The Trade Union which represents employees in the electrotechnical industry in the new millennium is called Amicus

◆ Through a network of local area offices the Union offers advice and support for its members. They will also provide legal advice and representation if a member has a serious accident as a result of a Health and Safety issue or has a dispute with an employer

◆ Further information can be found at www.amicustheunion.org

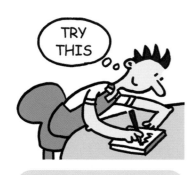

Does the Company you work for belong to a trade organisation?

Why do they belong, what are the advantages?

Do you belong to a Trade Union – if so, which one?

If not, why? Trade Union membership is often free while you are training.

Communications and Technical Information

When we talk about good communications we are talking about transferring information from one person to another both quickly and accurately. We do this by talking to other people, looking at drawings and plans and discussing these with colleagues from the same company and with other professionals who have an interest in the same project. The technical information used within our industry comes from many sources. The IEE Regulations (BS 7671) is the 'electricians' bible' and forms the basis of all our electrical design calculations and installation methods. British Standards, European Harmonised Standards and Codes of Practice provide detailed information for every sector of the electrotechnical industry, influencing all design and build considerations.

Equipment and accessories available to use in a specific situation can often be found in the very comprehensive manufacturers' catalogues and the catalogues of the major Wholesalers that service the Electrotechnical Industries.

All of this technical information may be distributed and retrieved by using:

- conventional drawings and diagrams which we will look at in more detail below

- sketch drawing to illustrate an idea or the shape of say a bracket to hold a piece of electrical equipment

- the Internet can be used to download British Standards and Codes of Practice

◆ the Internet can also be used to download Health and Safety information from the Health & Safety Executive at www.gov.uk/hse or www.opsi.gov.uk

◆ the Facsimile (Fax) machine and E.mail can be used to communicate with other busy professionals, information say about a project you are working on together

Let us now look at the types of drawings and diagrams which we use within our industry to communicate technical information between colleagues and other professionals. The type of diagram to be used in any particular situation is the one which most clearly communicates the desired information.

Site Plans or Layout Drawings

These are scale drawings based upon the architect's site plan of the building and show the position of the electrical equipment which is to be installed. The electrical equipment is identified by a graphical symbol. The standard symbols used by the electrical contracting industry are those recommended by the British Standard EN 60617, *Graphical Symbols for Electrical Power, Telecommunications and Electronic Diagrams*. Some of the more common electrical installation symbols are given in Fig. 1.13.

The Layout drawing or site plan of a small domestic extension is shown in Fig. 1.14. It can be seen that the mains intake position, probably a Consumer Unit, is situated in the store-room which also contains one light controlled by a switch at the door. The bathroom contains one lighting point controlled by a one-way pull switch at

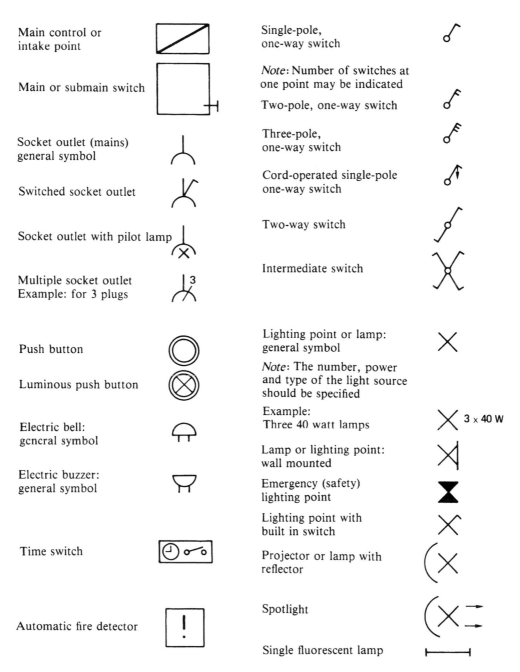

Figure 1.13 Some BS EN 60617 electrical installation symbols

Figure 1.14   Layout drawing or site plan of a small electrical installation

the door. The kitchen has two doors and a switch is installed at each door to control the fluorescent lumi-naire. There are also three double sockets situated around the kitchen. The sitting room has a two-way switch at each door controlling the centre lighting point. Two wall lights with built-in switches are to be wired, one at each side of the window. Two double sockets and one switched socket are also to be installed in the sitting room. The bedroom has two lighting points controlled independently by two one-way switches at the door. The wiring diagrams and installation procedures for all these circuits can be found in the next chapter.

As-fitted Drawings

When the installation is completed a set of drawings should be produced which indicate the final positions of all the electrical equipment. As the building and electrical installation progresses, it is sometimes necessary to modify the positions of equipment indicated on the layout drawing because, for example, the position of a doorway has been changed. The layout drawings or site plans indicate the original intentions for the position of equipment, while the 'as-fitted' drawing indicates the actual positions of equipment upon completion of the contract.

Detail Drawings and Assembly Drawings

These are additional drawings produced by the architect to clarify some point of detail. For example, a drawing might be produced to give a fuller description of a suspended ceiling arrangement or the assembly arrangements of the metalwork for the suspended ceiling.

Location Drawings

Location drawings identify the place where something is located. It might be the position of the manhole covers giving access to the drains. It might be the position of all water stop taps or the position of the emergency lighting fittings. This type of information may be placed on a blank copy of the architect's site plan or on a supplementary drawing.

Distribution Cable Route Plans

On large installations there may be more than one position for the electrical supplies. Distribution cables may radiate from the site of the electrical mains intake position to other sub-mains positions. The site of the sub-mains and the route taken by the distribution cables may be shown on a blank copy of the architect's site plan or on the electricians 'As-fitted' drawings.

Block Diagrams

A block diagram is a very simple diagram in which the various items or pieces of equipment are represented by a square or rectangular box. The purpose of the block diagram is to show how the components of the circuit relate to each other and, therefore, the individual circuit connections are not shown. Figure 1.15 shows the block diagram of a space heating control system.

Figure 1.15 Block diagram – space heating control system (Honeywell Y. Plan)

Wiring Diagrams

A wiring diagram or connection diagram shows the detailed connections between components or items of equipment. They do not indicate how a piece of equipment or circuit works. The purpose of a wiring diagram is to help someone with the actual wiring of the circuit. Figure 1.16 shows the wiring diagram for a space heating control system. Other wiring diagrams can be seen in Figs 4.8 and 4.9 (see Chapter 4).

Figure 1.16 Wiring diagram – space heating control system (Honeywell Y. Plan)

Circuit Diagrams

A circuit diagram shows most clearly how a circuit works. All the essential parts and connections are represented by their graphical symbols. The purpose of a circuit diagram is to help our understanding of the circuit. It will be laid out as clearly as possible, without regard to the physical layout of the actual components and, therefore, it may not indicate the most convenient way to wire the circuit. Figure 1.17 shows the circuit diagram of our same space heating control system. Figs 2.4 and 2.5 in Chapter 2 are circuit diagrams.

Figure 1.17   Circuit diagram – space heating control system (Honeywell Y. Plan)

Schematic Diagrams

A schematic diagram is a diagram in outline of, for example, a motor starter circuit. It uses graphical symbols to indicate the inter-relationship of the electrical elements in a circuit. These help us to understand the working operation of the circuit but are not helpful in showing us how to wire the components. An electrical schematic diagram looks very like a circuit diagram. Figure 1.18 shows a schematic diagram.

Figure 1.18 Schematic diagram – DOL motor starter

C1	Hold-in contactor coil
2	Main contacts
C3	Retaining contact
4	Start/close button
5	Stop/open button
6	Magnetic or thermal overload trip coils
7	Overload trip contact

Freehand Working Diagrams

Freehand working drawings or sketches are another important way in which we communicate our ideas. The drawings of the spring toggle bolt in Chapter 4 (Fig. 4.34) were done from freehand sketches. A freehand sketch may be done as an initial draft of an idea before a full working drawing is made. It is often much easier to produce a sketch of your ideas or intentions than to describe them or produce a list of instructions.

To convey the message or information clearly it is better to make your sketch large rather than too small. It should also contain all the dimensions necessary to indicate clearly the size of the finished object depicted by the sketch.

If you have read and understood the whole of this Chapter, you have completed all of the underpinning knowledge requirements of the first core unit in the City & Guilds 2330 Syllabus for the Level 2 Certificate in Electrotechnical Technology.

When you have completed the practical assessments required by the City & Guilds Syllabus, which you are probably doing at your local college, you will be ready to tackle the on-line assessment.

The Positional Reference System

A positional reference system can be used to mark exact positions in any space. It uses a simple grid reference system to mark out points in the space enclosed by the grid. It is easy to understand if we consider a specific example which I use when building prototype electronic circuits on matrix board. Matrix board is the insulated board full of holes into which we insert small pins and then attach the electronic components.

To set up the grid reference, count along the columns at the top of the board, starting from the left and then count down the rows. The position of point 4:3 would be 4 holes from the left and 3 holes down.

Prepare a Matrix board, or any space for that matter, as follows:

◆ Turn the Matrix board so that a manufactured straight edge is to the top and left-hand side

◆ Use a felt tip pen to mark the holes in groups of five along the top edge and down the left-hand edge as shown in Fig. 1.19

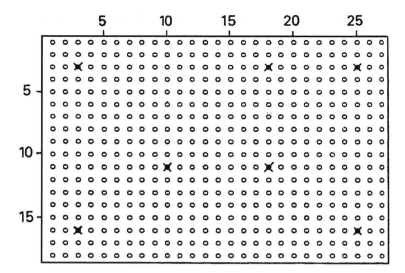

Figure 1.19 Positional reference system used to identify points on an electronic matrix board

So, to prepare you for the multiple choice on-line assessment, try the following multiple choice questions.

◆ The pins can then be inserted as required. Figure 1.19 shows a number of pin reference points. Counting from the left-hand side of the board there are 3:3, 3:16, 10:11, 18:3, 18:11, 25:3 and 25:16.

Assessment Questions

Identify the statements as true or false. If only part of the statement is false, tick false

1   The Health & Safety at Work Act provides the legal framework for encouraging good standards of health and safety at work.
True ❑   False ❑

2   The Wiring Regulations (BS 7671) are recognised as the National Standard for electrical work in the UK.
True ❑   False ❑

3   The Health & Safety at Work Act is a Statutory Regulation.
True ❑   False ❑

4   The Wiring Regulations (BS 7671) are Statutory Regulations.
True ❑   False ❑

5   An employee – that is a worker – is responsible for his or her own safety at work and the safety of others who might be affected by what they do.
True ❑   False ❑

6   There are four types of safety signs which inform workers of the rules and regulations especially relevant to a particular work situation.

Warning Signs describe what <u>must not</u> be done.
True ❑   False ❑

7 Mandatory Signs give information which <u>must be obeyed</u> in the work environment.
True ❏   False ❏

8 Site plans or layout drawings are scaled drawings showing the position of the electrical equipment to be installed.
True ❏   False ❏

9 Block diagrams show the detailed connections between components or pieces of equipment.
True ❏   False ❏

10 The ECA is a trade organisation representing electrotechnical companies and AMICUS is a Trade Union representing employees.
True ❏   False ❏

Multiple Choice Assessment Questions

Tick the correct answer. Note that more than ONE answer may be correct

11 **Identify the Regulations which are Statutory Regulations**
    a   The HSAWA (Health & Safety at Work Act)   ❑
    b   E @ Work Regs. (Electricity at Work Regulations)   ❑
    c   COSHH (Control of Substances Hazardous to Health)   ❑
    d   IEE Wiring Regs. (BS 7671)   ❑

12 **Identify the Regulations which are Non-Statutory Regulations**
    a   The HSAWA (Health & Safety at Work Act)   ❑
    b   E @ Work Regs. (Electricity at Work Regulations)   ❑
    c   COSHH (Control of Substances Hazardous to Health)   ❑
    d   IEE Wiring Regs. (BS 7671)   ❑

13 **The HSAWA puts the responsibility for safety at work upon:**
    a   an employee   ❑
    b   an employer   ❑
    c   everyone   ❑
    d   the Government   ❑

14 **To work safely and care for the safety of others is the responsibility of:**
    a   an employee   ❑
    b   an employer   ❑

c everyone ☐

d the Government ☐

15 To prepare a Health & Safety Policy Statement is the responsibility of:

a an employee ☐

b an employer ☐

c everyone ☐

d the Government ☐

16 Triangular yellow safety signs with a black border and symbol are called:

a Advisory signs giving safety information ☐

b Mandatory signs or **must do** signs ☐

c Prohibition signs or **must not do** signs ☐

d Warning signs giving safety information ☐

17 Square or rectangular green signs with a white symbol are called:

a Advisory signs giving safety information ☐

b Mandatory signs or **must do** signs ☐

c Prohibition signs or **must not do** signs ☐

d Warning signs giving safety information ☐

18 Circular blue signs with a white symbol are called:

a Advisory signs giving safety information ☐

b Mandatory signs or **must do** signs ☐

c Prohibition signs or **must not do** signs ☐

d Warning signs giving safety information ☐

19 Circular white signs with a red border and red cross bar are called:

a Advisory signs giving safety information ☐

b Mandatory signs or **must do** signs ☐

   c Prohibition signs or **must not do** signs ❑

   d Warning signs giving safety information ❑

20 **An uncontrolled event causing injury or damage is one definition of:**

   a a runaway bus ❑

   b a first aid procedure ❑

   c an accident ❑

   d an emergency procedure ❑

21 **A fire extinguisher showing a signal red flash on a red background contains:**

   a Carbon dioxide gas ❑

   b Dry powder ❑

   c Foam ❑

   d Water ❑

22 **A fire extinguisher showing a black flash on a red background contains:**

   a Carbon dioxide gas ❑

   b Dry powder ❑

   c Foam ❑

   d Water ❑

23 **A fire extinguisher showing a pale cream flash on a red background contains:**

   a Carbon dioxide gas ❑

   b Dry powder ❑

   c Foam ❑

   d Water ❑

24 **Following every accident at work:**

   a an employee must take three days off work ❑

b a waterproof plaster must be placed on
the injury ☐

c a record must be made in the Accident/
First Aid book ☐

d a report must be sent to the HSE local
area office ☐

25 **The Electricity at Work Regulations tell us
that before work commences on electrical
equipment it must be disconnected
from the source of supply and that
disconnection must be secure. To comply
with this Regulation we must:**

a switch off the circuit at the local functional
switch ☐

b switch off the current at the local isolator
switch ☐

c follow a suitable electrical isolation
procedure ☐

d follow the test procedures given in Part 7
of the IEE Regulations (BS 7671) ☐

26 **Emergency lighting and security systems
ensure that a building:**

a is safe to use in unforeseen circumstances ☐

b is illuminated and heated to an
appropriate level ☐

c ensures the efficiency of the installed system ☐

d provides safe monitoring of industrial
processes and systems ☐

27 **Electrical maintenance**

a is safe to use in unforeseen circumstances ☐

b is illuminated and heated to an
appropriate level ☐

    c ensures the efficiency of the installed
       system ❏

    d provides safe monitoring of industrial
       processes and systems ❏

28 The Supervisor/Foreman will:

    a oversee a number of electrical contracts ❏

    b be responsible for the day to day
       management of one specific contract ❏

    c be the leader of a small team installing
       electrical systems ❏

    d be an operative who has a basic skill
       and qualification in mechanical rather
       than electrical engineering ❏

29 The Contracts Manager of a Company will:

    a oversee a number of electrical contracts ❏

    b be responsible for the day to day
       management of one specific contract ❏

    c be the leader of a small team installing
       electrical systems ❏

    d be an operative who has a basic skill and
       qualification in mechanical rather than
       electrical engineering ❏

30 A Mechanic/Fitter will:

    a oversee a number of electrical contracts ❏

    b be responsible for the day to day
       management of one specific contract ❏

    c be the leader of a small team installing
       electrical systems ❏

d be an operative who has a basic skill and qualification in mechanical rather than electrical engineering ☐

31 A Trade Union is:

a the electrical industry's safety regulatory body ☐

b a professional body supporting electrotechnical organisations ☐

c the British Standard for electrical power supplies ☐

d an organisation representing electrical employees ☐

32 The National Inspection Council for Electrical Installation contracting is:

a the electrical industry's safety regulatory body ☐

b a professional body supporting electrotechnical organisations ☐

c the British Standard for electrical power supplies ☐

d the Trade Union representing electrical employees ☐

33 A scale drawing showing the position of equipment by graphical symbol is a description of a:

a block diagram ☐

b layout diagram or site plan ☐

c wiring diagram ☐

d circuit diagram ☐

34 **A diagram which shows the detailed connection between individual items of equipment is a description of a:**
  a   block diagram ❑
  b   layout diagram or site plan ❑
  c   wiring diagram ❑
  d   circuit diagram ❑

35 **A diagram which shows most clearly how a circuit works, with all items represented by graphical symbols is a description of:**
  a   block diagram ❑
  b   layout diagram or site plan ❑
  c   wiring diagram ❑
  d   circuit diagram ❑

Basic Principles of Electrotechnology

Chapter 2 covers the topics described in the second core unit of the City & Guilds 2330 Syllabus for the Level 2 Certificate in Electrotechnical Technology

This Chapter describes the basic scientific concepts and electrical circuits which form the foundations of electrotechnology.

Basic Units used in Electrotechnology

In all branches of science, engineering and electro-technology we use the international metric system of units called the System International, abbreviated to SI system.

Table 2.1 describes some of the basic units that we shall be using in this chapter.

Table 2.1   Basic SI Units

Quantity	Measure of	Basic Unit	Symbol	Notes
area	length × length	metre squared	m^2	
current I	electric current	ampere	A	
energy	ability to do work	Joule	J	Joule is a very small unit 3.6×10^6 J = 1 kWh
force	the effect on a body	Newton	N	
frequency	number of cycles	Hertz	Hz	mains frequency is 50 Hz
length	distance	metre	m	
mass	amount of material	kilogram	kg	1 metric tonne = 1000 kg
magnetic flux Φ	magnetic energy	Weber	Wb	
magnetic flux density B	number of lines of magnetic flux	Tesla	T	

Table 2.1   (*Continued*)

Quantity	Measure of	Basic Unit	Symbol	Notes
potential or pressure	voltage	volt	V	
period T	time taken to complete one cycle	second	s	the 50 Hz mains supply has a period of 20 ms
power	rate of doing work	Watt	W	
resistance	opposition to current flow	Ohm	Ω	
resistivity	resistance of a sample piece of material	Ohm metre	ρ	resistivity of copper is $17.5 \times 10^{-9}\,\Omega\text{m}$
temperature	hotness or coldness	Kelvin	K	0°C = 273 K. A change of 1 K is the same as 1°C
time	time	second	s	60 s = 1 min 60 min = 1 h
weight	force exerted by a mass	kilogram	kg	1000 kg = 1 tonne

Note: A more detailed description can be found in Unit 2 of *Basic Electrical Installation Work* 4th Edition, ISBN 0750666242.

Like all metric systems, SI units may be increased or reduced by using multiples or sub-multiples of 10. Some of the more common multiples and their names are shown in Table 2.2.

The unit of electrical power is the watt, symbol W, but this is a small unit of power and a more common unit

Table 2.2 Symbols and multiples for use with SI units

Prefix	Symbol	Multiplication factor		
Mega	M	$\times 10^{6}$	or	$\times 1000000$
Kilo	k	$\times 10^{3}$	or	$\times 1000$
Hecto	h	$\times 10^{2}$	or	$\times 100$
Decca	da	$\times 10$	or	$\times 10$
Deci	d	$\times 10^{-1}$	or	$\div 10$
Centi	c	$\times 10^{-2}$	or	$\div 100$
Milli	m	$\times 10^{-3}$	or	$\div 1000$
Micro	m	$\times 10^{-6}$	or	$\div 1000000$

TRY THIS

Can you think of other multiples and sub-multiples of basic units? Ask your workmates, work supervisor or College Lecturer.

is the kilowatt or one thousand watts. This is expressed as kW in the SI system of units.

Electrical Theory

◆ All matter is made up of atoms.

◆ All atoms are made up of a central positively charged nucleus surrounded by negatively charged electrons.

◆ The electrical properties of materials depends largely upon how tightly these electrons are bound to the nucleus.

◆ **A conductor** is a material in which the electrons are loosely bound to the central nucleus and, in fact, can very easily become free electrons. These free electrons drift around randomly inside a conductor as shown in Fig. 2.1(a).

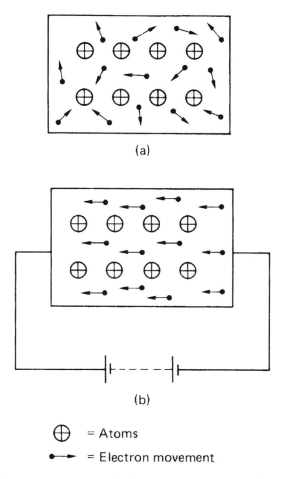

(a)

(b)

⊕ = Atoms

●→ = Electron movement

Figure 2.1 Atoms and electrons in a material.
(a) Shows the random movement of free electrons;
(b) shows the free electrons drifting towards the
positive terminal when a voltage is applied

◆ Good conductors are gold, silver, copper,
aluminium, brass etc.

◆ **An insulator** is a material in which the
electrons are very tightly or strongly bound to
the central nucleus.

◆ Good insulators are PVC, rubber, perspex, glass,
wood, porcelain etc.

Electron Flow or Electric Current

◆ If a battery is attached to a 'good conductor' material, the free electrons drift toward the positive terminal as shown in Fig. 2.1b.

◆ The drift of electrons within a conductor is what we know as an electric current flow.

◆ Current flow is given the symbol I and is measured in amperes.

Electrical Cables

Electrical cables are used to carry electric currents.

Most cables are constructed in three parts:

1 The conductor, that carries the current and may have a stranded or solid core.

2 The insulation, that contains the current and is colour coded for identification.

3 The outer sheath that may contain some means of providing protection from mechanical damage.

Figure 2.2 shows a PVC insulated and sheathed cable. The type used for domestic installations.

Figure 2.3 shows a PVC/SWA (PVC insulated steel wire armoured) cable. The type used for industrial or underground installations where some mechanical protection is required.

Figure 2.2   A twin and earth PVC insulated and sheathed cable

Figure 2.3   A four core PVC/SWA cable

Three Effects of an Electric Current

When an electric current flows in a circuit it can have one or more of the following three effects: **heating**, **magnetic** or **chemical**.

Heating Effect

◆   The electrons moving in the conductor causes the conductor to heat up

◆ The amount of heat generated depends upon the:
1 amount of current flowing
2 dimensions of the conductors
3 type of conductor material used

◆ Practical applications of the heating effect of an electric current are:
1 radiant heaters which heat rooms
2 circuit protection fuses and MCBs which cut off the supply when an overcurrent flows

Magnetic Effect

◆ Whenever a current flows in a conductor a magnetic field is set up around the conductor like an extension of the insulation – more about this later

◆ Increasing the current increases the magnetic field

◆ Switching the current off causes the magnetic field to collapse

◆ Practical applications of the magnetic effect are:
1 electric motors which rotate because of the magnetic flux generated by the electrical supply door chimes and buzzers which ding dong or buzz because of the magnetic flux generated by the electrical supply

Chemical Effect

◆ When an electric current flows through a conducting liquid, the liquid separates into

its chemical parts, a process called
electrolysis

◆   Alternatively, if two metals are placed in a
conducting liquid they react chemically and
produce a voltage

◆   Practical applications of the chemical effect
are:
1   industrial processes such as electroplating
which is used to silver plate sports trophies
and cutlery
2   motor car batteries which store electrical
energy

Ohm's Law

This is one of the most famous electrical laws published by Dr George Ohm in 1826. It allows us to
understand the relationship between the basic elements of an electric circuit, voltage current and resistance. Voltage is the pressure or potential driving
current around a circuit. Current, as we saw a little
earlier at Fig. 2.1 is the movement of electrons
through a conductor and resistance is the opposition
to that current flow. His law may be expressed as
voltage is equal to current times resistance or
expressed mathematically as:

$$V = I \times R \text{ volts}$$

Transposing this formula, we have:

$$\text{Current } I = \frac{V}{R} \text{ (A)} \quad \text{and} \quad \text{Resistance } R = \frac{V}{I} \text{ (}\Omega\text{)}$$

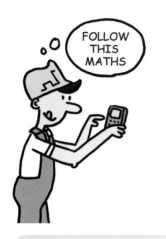

Follow this Maths carefully step by step

EXAMPLE 1

An electric fan heater was found to take **10 A** when connected to the **230 V** mains supply. Calculate the resistance of the heater element.

$$\text{From } R = \frac{V}{I} \, (\Omega)$$

$$R = \frac{230V}{10A} = 23 \, (\Omega)$$

The heater element resistance is 23 ohm

EXAMPLE 2

Calculate the current flowing in a disco 'sound and light' unit having a resistance of **57.5 Ω** when it is connected to the **230 V** electrical mains.

$$\text{From } I = \frac{V}{R} \, (A)$$

$$I = \frac{230V}{57.5\Omega} = 4 \, (A)$$

The 'sound and light' unit takes 4 amps

Resistivity

The resistance or opposition to current flow varies, depending upon the type of material being used to carry the electric current.

Resistivity is defined as the resistance of a sample of a particular material and Table 2.3 gives the resistivity values of some common materials.

Table 2.3 Resistivity values	
Material	**Resistivity (Ωm)**
Silver	16.4×10^{-9}
Copper	17.5×10^{-9}
Aluminium	28.5×10^{-9}
Brass	75.0×10^{-9}
Iron	100.0×10^{-9}

Using these values we can calculate the resistance of different materials using the formulae

$$\text{Resistance } R = \frac{\rho l}{a} \; (\Omega)$$

where ρ (the Greek letter rho) is the resistivity value for the material, l is the length and a is the cross-sectional area

FOLLOW THIS MATHS

EXAMPLE 3

Calculate the resistance of **100 m** of **2.5 mm²** copper cable using the resistivity values in Table 2.3.

$$\text{We know that } R = \frac{\rho l}{a} \; \Omega$$

$$\text{therefore } R = \frac{17.5 \times 10^{-9} \times 100}{2.5 \times 10^{-6}}$$

$$\therefore R = 700 \times 10^{-3} \; (\Omega) \text{ or}$$

$$R = 700 \; (\text{m}\Omega)$$

Follow this Maths carefully step by step

Note: the cross section of the cable is in mm²
mm = 10⁻³ (see Table 2.2) so,
mm × mm = 10⁻⁶

EXAMPLE 4

Calculate the resistance of **100 m** of **2.5 mm²** aluminium cable, using the resistivity values in Table 2.3.

$$R = \frac{\rho l}{a} \ (\Omega)$$

$$\text{Therefore } R = \frac{28.5 \times 10^{-9} \times 100}{2.5 \times 10^{-6}}$$

$$\therefore R = 1140 \times 10^{-3} \ (\Omega) \text{ or}$$
$$R = 1140 \ (m\Omega)$$

Series Connected Resistors

When resistors are connected as shown in Fig. 2.4 we say they are connected in series. The same current flows through each resistor and so we say the current is 'common'. When the current flows through R_1 there will be a volt drop across R_1 because of ohm's law

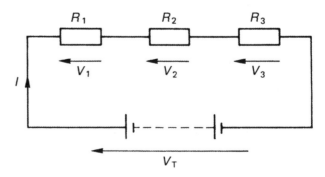

Figure 2.4   A series circuit

$V = I \times R$. For the same reason a volt drop will occur across R_2 and R_3. The addition of the three volt drops will add up to the total voltage V_T

$$\text{so, } V_T = V_1 + V_2 + V_3 \text{ volts}$$

and from the calculations made in Ohm's Law:

$$\text{Total Resistance } R_T = R_1 + R_2 + R_3 \text{ ohms}$$

The unit of resistance is the ohm to commemorate the great work done by Dr George Ohm.

Parallel Connected Resistors

When resistors are connected as shown in Fig. 2.5 we say they are connected in parallel. The same voltage is connected across each resistor and so we say the voltage is common in a parallel circuit. When the current reaches the resistor junction, it will divide, part of it flowing through each resistor. The addition

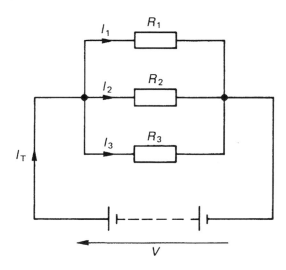

Figure 2.5 A parallel circuit

Series and Parallel Resistors and the different formulae used is important information to be remembered of the three currents will add up to the total current drawn from the battery, so:

$$I_T = I_1 + I_2 + I_3 \text{ amps}$$

and from the calculations made in Ohm's Law:

Total resistance is found from $\dfrac{1}{R_T} = \dfrac{1}{R_1} + \dfrac{1}{R_2} + \dfrac{1}{R_3}$

EXAMPLE 5

Three $6\,\Omega$ resistors are connected (a) in series (see Fig. 2.6), and (b) in parallel (see Fig. 2.7), across a 12 V battery. For each method of connection, find the total resistance and the values of all currents and voltages.

For any series connection:

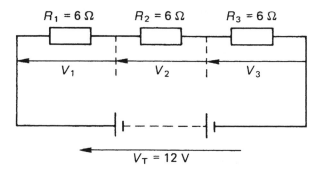

Figure 2.6   Resistors in series

$$R_T = R_1 + R_2 + R_3$$

$$\therefore R_T = 6\,\Omega + 6\,\Omega + 6\,\Omega = 18\,\Omega$$

Total current $I_T = \dfrac{V_T}{R_T}$

$$\therefore I_T = \dfrac{12\,V}{18\,\Omega} = 0.67\,A$$

The voltage drop across R_1 is

$$V_1 = I_T \times R_1$$
$$\therefore V_1 = 0.67\,\text{A} \times 6\,\Omega = 4\,\text{V}$$

The voltage drop across R_2 is

$$V_2 = I_T \times R_2$$
$$\therefore V_2 = 0.67\,\text{A} \times 6\,\Omega = 4\,\text{V}$$

The voltage drop across R_3 is

$$V_3 = I_T \times R_3$$
$$\therefore V_3 = 0.67\,\text{A} \times 6\,\Omega = 4\,\text{V}$$

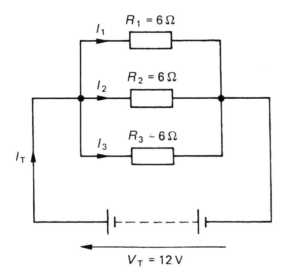

Figure 2.7   Resistors in parallel

For any parallel connection,

$$\frac{1}{R_T} = \frac{1}{R_1} + \frac{1}{R_2} + \frac{1}{R_3}$$

$$\therefore \frac{1}{R_T} = \frac{1}{6\,\Omega} + \frac{1}{6\,\Omega} + \frac{1}{6\,\Omega}$$

$$\frac{1}{R_T} = \frac{1+1+1}{6\,\Omega} = \frac{3}{6\,\Omega}$$

$$R_T = \frac{6\,\Omega}{3} = 2\,\Omega$$

$$\text{Total current } I_T = \frac{V_T}{R_T}$$

$$\therefore I_T = \frac{12\,V}{2\,\Omega} = 6\,A$$

The current flowing through R_1 is

$$I_1 = \frac{V_T}{R_1}$$

$$\therefore I_1 = \frac{12\,V}{6\,\Omega} = 2\,A$$

The current flowing through R_2 is

$$I_2 = \frac{V_T}{R_2}$$

$$\therefore I_2 = \frac{12\,V}{6\,\Omega} = 2\,A$$

The current flowing through R_3 is

$$I_3 = \frac{V_T}{R_3}$$

$$\therefore I_3 = \frac{12\,V}{6\,\Omega} = 2\,A$$

Component Parts of an Electrical Circuit

These series and parallel resistors are connected together to form an electrical circuit. So, what is an electrical circuit?

An electrical circuit has the following **five** components:

◆ a source of electrical energy. This might be a battery giving a D.C. (direct current) supply or the mains supply which is A.C. (alternating current)

◆ a source of circuit protection. This might be a fuse or circuit breaker which will protect the circuit from 'overcurrent'

◆ the circuit conductors or cables. These carry voltage and current to power the load

◆ a means to control the circuit. This might be a simple on/off switch but it might also be a dimmer or a thermostat

◆ and a load. This is something which needs electricity to make it work. It might be an electric lamp, an electrical appliance, an electric motor or an i-pod

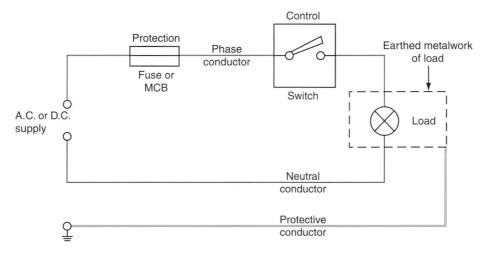

Figure 2.8   Component parts of an electric circuit

Connecting Voltmeters and Ammeters

From the work discussed before we now know that current flows **through** a conductor and voltage appears **across** a resistor, a lamp or any other load. And so, this gives us a good indication of how to connect a voltmeter or ammeter.

An ammeter must have the current flowing through it and so is connected in series with the load. A voltmeter must be connected across the load and so is connected in parallel with the load.

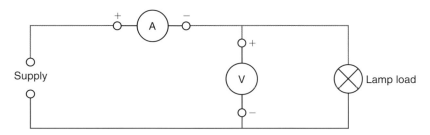

Figure 2.9   Connecting voltmeters and ammeters

Figure 2.9 shows a voltmeter and ammeter connected to measure the current and voltage in a lamp load.

Magnetic Fields and Flux Patterns

Lines of magnetic flux have no physical existence but were introduced by Michael Faraday as a way of explaining the magnetic energy existing in space or in a material. The magnetic fields around a permanent magnet, a current carrying conductor and a solenoid are shown in Figs 2.10, 2.11 and 2.12.

Bar magnet

Horse shoe magnet

Figure 2.10 Magnetic field around a permanent magnet

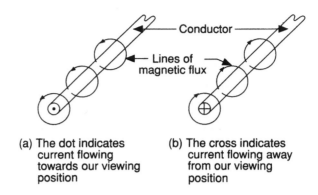

(a) The dot indicates current flowing towards our viewing position

(b) The cross indicates current flowing away from our viewing position

Figure 2.11   Magnetic field around a current carrying relay

Lines of magnetic flux around a solenoid with no core

Lines of magnetic flux around a solenoid with a soft iron core

Simple relay

Figure 2.12   The solenoid and one practical application, the relay

It is magnetic energy which is used to make commercial electricity today. This was Michael Faraday's great discovery in 1831. Magnetic energy turns electric motors and drives the wheels of industry and is, therefore, important in electrotechnology.

Basic Mechanics and Machines

Mechanics is the scientific study of machines, where a machine may be defined as any device which transmits motion from one place to another. So a lever, a wheel and axle and a pulley are all basic machines. A modern car engine is an energy transforming machine converting fuel energy into motion.

The City & Guilds syllabus asks us to consider weight, mass, force and work done by a force so let us define some of these scientific terms.

Mass   This is a measure of the amount of material in a substance such as wood or metals

Weight   This is a measure of the force which the mass exerts. It exerts this force because it is being attracted towards the earth by gravity

Force   The presence of a force can only be detected by its effect on an object. A force may cause a stationary object to move or a moving object to stop.

Gravity   The force of gravity acts toward the centre of the earth and causes objects to fall to the ground at a rate of 9.81 m/s

Work done   The work done by a force is a measure of the force exerted times the distance moved in the direction of the force

Suppose a broken-down motor car was to be pushed along a road; work would be done on the car by applying the force necessary to move it along the road. Heavy breathing and perspiration would be evidence of the work done:

Work done = Force × Distance moved in the
direction of the force (J)

The SI unit of work done is the newton metre or joule (symbol J). The joule is the preferred unit and it commemorates an English physicist, James Prescot Joule (1818–89).

FOLLOW THIS MATHS

Follow this Maths carefully step by step so that you understand where the numbers come from

EXAMPLE 6

A building hoist lifts ten 50 kg bags of cement through a vertical distance of 30 m to the top of a high rise building. Calculate the work done by the hoist, assuming the acceleration due to gravity to be 9.81 m/s^2.

$$\text{Work done} = \text{Force} \times \text{Distance moved (J)}$$
$$\text{but} \quad \text{Force} = \text{Mass} \times \text{Acceleration (N)}$$
$$\therefore \text{Work done} = \text{Mass} \times \text{Acceleration}$$
$$\times \text{Distance moved (J)}$$
$$\text{Work done} = 10 \times 50\,\text{kg} \times 9.81\,\text{m/s}^2 \times 30\,\text{m}$$
$$\text{Work done} = 147.15\,\text{kJ}.$$

Power

If one motor car can cover the distance between two points more quickly than another car, we say that the faster car is more powerful. It can do a given amount of work more quickly. By definition, power is the rate of doing work.

$$\text{Power} = \frac{\text{Work done}}{\text{Time taken}}\,(\text{W})$$

The SI unit of power, both electrical and mechanical, is the watt (symbol W). This commemorates the name of James Watt (1736–1819), the inventor of the steam engine.

EXAMPLE 7

A building hoist lifts ten 50 kg bags of cement to the top of a 30 m high building. Calculate the rating (power) of the motor to perform this task in 60 seconds if the acceleration due to gravity is taken as 9.81 m/s².

$$\text{Power} = \frac{\text{Work done}}{\text{Time taken}} \text{ (W).}$$

but   Work done = Force × Distance moved (J)
and   Force = Mass × Acceleration (N)

By substitution,

$$\text{Power} = \frac{\text{Mass} \times \text{Acceleration} \times \text{Distance moved}}{\text{Time taken}} \text{ (W)}$$

$$\text{Power} = \frac{10 \times 50\,\text{kg} \times 9.81\,\text{m/s}^2 \times 30\,\text{m}}{60\,\text{s}}$$

$$\text{Power} = 2452.5\,\text{W}$$

The rating of the building hoist motor will be 2.45 kW.

EXAMPLE 8

A hydroelectric power station pump motor working continuously during a 7 hour period raises 856 tonnes of water through a vertical distance of 60 m. Determine the rating (power) of the motor, assuming the acceleration due to gravity is 9.81 m/s².

From Example 7,

$$Power = \frac{Mass \times Acceleration \times Distance\ moved}{Time\ taken}\ (W)$$

$$Power = \frac{856 \times 1000\,kg \times 9.81m/s^2 \times 60\,m}{7 \times 60 \times 60\,s}$$

$$Power = 20\,000\,W$$

The rating of the pump motor is 20 kW.

EXAMPLE 9

An electric hoist motor raises a load of 500 kg at a velocity of 2 m/s. Calculate the rating (power) of the motor if the acceleration due to gravity is 9.81 m/s^2.

$$Power = \frac{Mass \times Acceleration \times Distance\ moved}{Time\ taken}\ (W)$$

$$but \quad Velocity = \frac{Distance}{Time}\ (m/s)$$

$$\therefore Power = Mass \times Acceleration \times Velocity$$

$$Power = 500\ kg \times 9.81m/s^2 \times 2\ m/s$$

$$Power = 9810\ W.$$

The rating of the hoist motor is 9.81 kW.

Efficiency

In any machine the power available at the output is less than that which is put in because losses occur in the machine. The losses may result from friction in the bearings, wind resistance to moving parts, heat, noise or vibration.

The ratio of the output power to the input power is known as the *efficiency* of the machine. The symbol for efficiency is the Greek letter 'eta' (η). In general,

$$\eta = \frac{\text{Power output}}{\text{Power input}}$$

Since efficiency is usually expressed as a percentage we modify the general formula as follows.

$$\eta = \frac{\text{Power output}}{\text{Power input}} \times 100$$

EXAMPLE 10

A transformer feeds the 9.81 kW motor driving the mechanical hoist of the previous example. The input power to the transformer was found to be 10.9 kW. Find the efficiency of the transformer.

$$\eta = \frac{\text{Power output}}{\text{Power input}} \times 100$$

$$\eta = \frac{9.81\,\text{kW}}{10.9\,\text{kW}} \times 100 = 90\%$$

Thus the transformer is 90% efficient. Note that efficiency has no units, but is simply expressed as a percentage.

The Simple Alternator

If a loop of wire is rotated between the poles of a magnet as shown in Fig. 2.13, the loop of wire will cut the lines of magnetic flux which pass from the north to the south pole. This flux cutting causes a voltage to be induced in the loop of wire. (Michael Faraday's Law)

If this induced voltage is collected by carbon brushes at the slip rings and displayed on a meter or CRO, it will be seen to induce first a positive and then a negative voltage. We call this changing voltage an alternating voltage and the shape that it follows is called, in mathematics, sinusoidal.

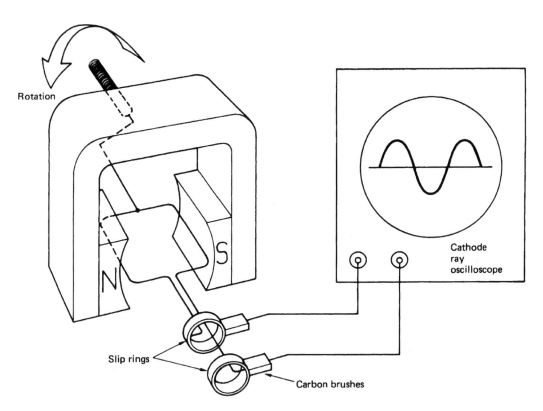

Figure 2.13   Simple A.C. generator or alternator

Electrical Transformers

A transformer is an electrical machine without moving parts, which is used to change the value of an alternating voltage.

A transformer will only work on an alternating supply, it will not normally work from a D.C. supply such as a battery.

Figure 2.14   A simple transformer

◆ A transformer such as that shown in Fig. 2.14 consists of two coils called the primary and secondary coils or windings, wound on to a common core. The iron core of the transformer is not solid but made up of very thin sheets called laminations, to improve efficiency.

◆ An alternating voltage applied to the primary winding establishes an alternating magnetic flux in the core.

◆ The magnetic flux in the core causes a voltage to be induced in the secondary winding of the transformers.

◆ The voltage in both the primary and secondary windings is proportional to the number of turns.

◆ This means that if you increase the number of secondary turns you will increase the output voltage. This has an application in power distribution.

◆ Alternatively, reducing the number of secondary turns will reduce the output voltage. This is useful for low voltage supplies such as domestic bell transformers. Because it has no moving parts, a transformer can have a very high efficiency. Large power transformers, used on electrical distribution systems, can have an efficiency of better than 90%.

These power transformers need cooling to take the heat generated away from the core. This is often achieved by totally immersing the core and windings in insulating oil. A sketch of an oil immersed transformer can be seen in Fig. 2.15.

Very small transformers are used in electronic applications. Small transformers are used as isolating transformers in shaver sockets and can also be used to supply SELV (separated extra low voltage) sources. Equipment supplied from a SELV source may be installed in a bathroom or shower-room, provided that it is suitably enclosed and protected from the ingress of moisture. This includes equipment such as water heaters, pumps for showers and whirlpool baths.

Figure 2.15   Typical oil filled power transformer

Electrical Power on the National Grid

Electricity is generated in large modern Power Stations at 25 kV (25,000 volts). It is then transformed up to 132 kV or 270 kV for transmission to other parts of the country on the National Grid network. This is a network of overhead conductors suspended on transmission towers which link together the Power Stations and the millions of users of electricity.

Raising the voltage to these very high values reduces the losses on the transmission network. 66 kV or 33 kV are used for secondary transmission lines and then these high voltages are reduced to 11 kV at local sub-stations for distribution to end users such as factories, shops and houses at 400 V and 230 V.

The ease and efficiency of changing the voltage levels is only possible because we generate an A.C. supply. Transformers are then used to change the voltage levels to those which are appropriate. Very high voltages for transmission, lower voltages for safe end use. This would not be possible if a D.C. supply was generated.

Figure 2.16 shows a simplified diagram of electricity distribution.

Figure 2.16   Simplified diagram of the distribution of electricity from power station to consumer

Safe Electrical Systems

Installing electrical systems which will be safe for those who will use them is absolutely fundamental to the safe use of electricity. Electrical systems installed in accordance with the IEE Regulations (BS 7671) will be safe for those who will use them.

Chapter 13 of the IEE Regulations tells us that where electrical equipment may become charged with electricity so as to cause a danger, any metalwork must be connected to earth. When we say connected to '**earth**' we mean the general conductive mass of the planet Earth, whose potential is taken as zero. '**Earthing**' is the act of connecting the '**exposed conductive parts**' of an installation to the main earthing terminal of the installation.

'**Exposed conductive parts**' are the metal parts of the installation which are not **normally live** but which may become live under fault conditions. For example, the metalwork of an electrical appliance or the trunking and conduits of the installation.

All other metalwork within a building is called '**extraneous conductive parts**' and this includes structural steelwork and other service pipes such as gas, water, radiators and sinks. The extraneous conductive parts are prevented from becoming live by '**bonding**' them together and connecting them to the main earthing terminal of the installation. The bonding process maintains an '**equipotential**' (of zero volts) between all exposed and extraneous conductive parts.

You should remember all of the definitions of words in this section. Perhaps you could make a list of the words and write down a short definition for each one. You will probably need to read this section more than once before you understand it, but it is important.

Principles of Electric Shock Protection

An electric shock occurs when a person becomes a part of the electrical circuit. We looked at electric shock in Chapter 1 at Fig. 1.8. The intensity of the electric shock will depend upon many factors such as age, fitness and the circumstances in which the shock is received. In general terms, a shock current of more than 50 mA can be fatal.

Electric shock may occur in two ways, through direct contact or indirect contact. **Direct contact** means actually touching live parts and we protect against direct contact by:

◆   insulating live parts

◆   placing barriers or enclosures around live parts

◆   placing obstacles in front of live parts

◆   placing live parts out of reach

Each of these methods keep people away from live electrical equipment.

Indirect contact means touching exposed conductive parts, such as the metalwork of an appliance, which has become live as a result of a fault. The potential voltage on this metalwork rises above earth potential and an electric shock may occur when someone touches the metalwork.

Various methods of protection against indirect contact are described in Section 413 of the IEE

Regulations (BS 7671) but the most universally used method for supplies in the UK is earthed equipotential bonding coupled with automatic disconnection of the supply.

Earthed Equipotential Bonding Coupled with Automatic Disconnection of the Supply

In the UK the most universally used method of protection against electric shock by indirect contact is earthed equipotential bonding coupled with automatic disconnection of the electrical supply. Earthed equipotential bonding was discussed at the beginning of this Section and is the process of connecting all exposed conductive parts and extraneous conductive parts to the main earthing terminal of the electrical installation. Automatic disconnection of the supply is achieved by fuses, MCBs and RCDs.

If the circuit shown earlier in Fig. 2.8 was operating normally, current would flow from the supply to the load along the phase conductor, through the load and back along the neutral conductor. The protective device would be chosen to carry this current. However, if a fault occurs, for example, a short circuit to earth between the phase conductor and the earthed metalwork of the load, current will flow from the supply to the load and then through the low resistance earthing and bonding of the installation back to the supply. This will cause a large current to flow and, in a healthy circuit, the protective device will operate very quickly to remove the danger.

Fuses, MCBs and RCDs provide earth fault protection, overload protection and short circuit protection where:

◆ a short circuit is a fault of negligible impedance (call it resistance for now) between live and neutral conductors

◆ an overload is a current which exceeds the rated value in an otherwise healthy circuit

In all cases the basic requirement for protection is that the fault current should be removed quickly and the circuit isolated. The IEE Regulations state that the protective device must operate very quickly to remove the danger and within a time of:

◆ 0.4 seconds for portable equipment supplied by socket outlet circuits

◆ 5.0 seconds for fixed equipment

◆ 0.2 seconds for construction sites, agricultural and horticultural premises

Electrical Tools and Equipment

Good quality, sharp tools are important to any craftsman, they enable learned skills to be used to the best advantage. The basic tools required by anyone in the electrotechnical industry are those used for stripping and connecting conductors. These are pliers, side cutters, a knife and an assortment of screwdrivers with flat bladed, Philips, crosshead, Pozidriv, Torx or Hexidriv bits. Figure 2.17 shows the basic hand tools required for making electrical connections.

Figure 2.17   The tools used for making electrical connections

The additional tools required by an electrical crafts-man will depend upon the type of electrotechnical work being undertaken. When wiring new houses or re-wiring old ones, the additional tools are those more associated with a bricklayer or carpenter and some examples are shown in Fig. 2.18.

When working on industrial installations, installing conduit, trunking and tray, the additional tools required by an electrician would more normally be those associated with a fitter or sheet metal fabricator and some examples are shown in Fig. 2.19.

The special tools required for cable tray bending, steel conduit bending and screw threading stocks

Safety rules for hand tools

◆ always use the correct tool for the job in hand and use it properly and sensibly

Figure 2.18   Some additional tools required by an electrician engaged in house wiring

Figure 2.19   Some additional tools required by an electrician engaged in industrial installations

Cable tray bending machine

Combined conduit vice and bending machine

MI crimping tool

Conduit stocks and dies: two views

Figure 2.20 Some special tools required by an electrician engaged in industrial installations

and dies, plus M.I. Cable crimping tools are shown in Fig. 2.20.

Electrical power tools reduce much of the hard work for any craftsman, allowing increase in productivity. Battery powered tools are very popular because they are very safe to use on site and are often now supplied with two battery packs so that while one is being used the other is on charge. Only 110 V power tools with leads are allowed on most construction sites these days. Figure 2.21 shows a selection of electrical power tools.

Before using any power tools, the craftsman should inspect the tool and any associated flexible cords for damage. If the power tool carries a PAT (portable

- always keep tools clean and sharp
- always keep tools in a toolbox and secure

Safety rules for power tools

- always check that the casing is not damaged
- always check that the cable is not damaged

Figure 2.21   Electrical power tools

◆ always check that the plug top is not damaged

◆ always check that no coloured conductors are showing anywhere on the flexible cord

appliance testing) label, a check should be made to ensure that the test date has not expired.

All tools are expensive and, therefore, attractive to a thief so, when not in use, all tools must be stored safely and securely.

Safe Working Practice

Every year thousands of people have accidents at their place of work despite the legal requirements laid down by the Health & Safety Executive. Many people recover quickly but an accident at work can result in permanent harm or even death.

At the very least, injuries hurt individuals. They may prevent you from doing the things you enjoy in your spare time and the result could be loss of earnings to you and loss of production and possibly damage to equipment for your employer. Your place of work may look harmless but it can be dangerous.

You have a responsibility under the Health & Safety at Work Act to:

◆ learn how to work safely and to follow company procedures of work

◆ obey all safety rules, notices and signs

◆ not interfere with or misuse anything provided for safety

◆ report anything that seems damaged, faulty or dangerous

◆ behave sensibly, not play practical jokes and not distract other people at work

◆ walk sensibly and not run around the workplace

◆ use the prescribed walkways

◆ drive only those vehicles for which you have been properly trained and passed the necessary test

◆ not wear jewellery which could become caught in moving parts if you are using machinery at work

◆ always check that mains power tools have been properly tested (PAT tested) and carry a label

Finally, and most importantly

◆ If it is broken or damaged in any way DO NOT USE IT

◆ Ask a 'competent person' (probably your Supervisor) to check it out

◆ Do not let anyone else use it

◆ always wear appropriate clothing and PPE if necessary

The principles laid down in the many Health & Safety at Work Regulations control our working environment and make our workplaces safer but despite all the legislation, workers continue to be injured and killed at work.

In the year 2004 the Health & Safety Executive (HSE) statistics show that 235 people died as a result of a work related injury. In addition, about 28,000 people have serious injuries each year and about 130,000 people each year receive minor work related injuries which result in an absence from work for more than three days.

The most common causes of accident at work are:

◆ slips, trips and falls from above ground. Safe working above ground is discussed at the beginning of Chapter 3

◆ manual handling, that is moving objects by hand which may result in strains, sprains and trap injury pains. Always use a mechanical aid to move heavy objects. Safe manual handling is discussed at the beginning of Chapter 3

◆ using equipment, machines and tools. Make sure your tools and equipment meet the safety rules described in the last section

◆ storing equipment badly which then becomes unstable and falls on someone

If you have read and understood the whole of this Chapter, you have completed all of the underpinning knowledge

Figure 2.22 Safe manual handling

◆ fire – we discussed fire safety in Chapter 1

◆ electricity – the safe use of electricity is what this book and our industry is about. Always use the 'safe isolation procedure' before work begins as described in Chapter 1

To help prevent Accidents at Work:

◆ always behave sensibly and responsibly

◆ keep your work area clean and tidy

◆ keep walkways clear

requirements of the **second** core unit in the City & Guilds 2330 syllabus for the Level 2 Certificate in Electrotechnical Technology.

When you have completed the practical assessments required by the City & Guilds syllabus, which you are probably doing at your local College, you will be ready to tackle the on-line assessment.

So, to prepare you for the on-line assessment, try the following Assessment Questions.

◆ clean up spills or wet patches on the floor

◆ screen off your work areas from the general public and other trades

◆ put tools and equipment away when not in use. Do not leave things lying around for others to fall over

◆ when working above ground level, use the good practice described in the next Chapter

◆ when moving objects by hand, use the good practice described in the next Chapter under the heading 'Safe manual handling'

Assessment Questions

Identify the statements as true or false. If only part of the statement is false, tick false

1 The electrical properties of any material basically depend upon how tightly the electrons are attached to the nucleus of the atom. A strong bond, and the material will be an insulator, a weak bond and the material will be a good conductor of electricity.
   True ❑   False ❑

2 An electrical current flowing in a conductor will have a heating, magnetic or chemical effect upon the circuit. Increasing the current flow will always **reduce** one of the three effects.
   True ❑   False ❑

3 The pioneering work carried out by Dr George Ohm in 1826 allows us today to calculate the relationship between current, voltage and resistance in an electric circuit.
   True ❑   False ❑

4 Measuring the current and voltage in an electric circuit is an important practical skill for anyone in the electrotechnical industry. When measuring current, the ammeter is always connected across the load. When measuring voltage, the voltmeter is always connected in series with the load.
   True ❑   False ❑

5 Electrical cables are used to carry electrical currents. Most cables are constructed in three parts:
- ◆ the conductor, which prevents human beings and livestock from an electric shock
- ◆ the insulation, which carries the electric current, thicker cables carry more current
- ◆ the outer sheath, which may incorporate a means of protection from mechanical damage

True ❏   False ❏

6 Good quality sharp tools are important to a craftsman in the electrotechnical industry. Always apply the following safety rules to both hand and power tools:
- ◆ keep tools sharp, clean and in a toolbox when not in use
- ◆ always use the correct tools and leads
- ◆ check for damage to power tools and leads
- ◆ mains power tools should be PAT tested and carry a label which is 'in date'
- ◆ if it's broken or damaged, don't use it, and don't let anyone else use it

True ❏   False ❏

7 Slips, trips and falls are the most common cause of accidents in the workplace. To prevent accidents at work always:
- ◆ behave sensibly and don't fool about
- ◆ keep your work area clean and avoid tripping hazards
- ◆ when working above ground, work from a suitable platform

◆ when lifting objects by hand, use the 'safe manual handling technique'

True ❑   False ❑

8 When those of us who work in the electrotechnical industry use the phrase 'to connect something to earth', we mean't connect that something to the main earthing terminal of the electrical installation so as to make it safe.

True ❑   False ❑

9 The metal trunking and conduits of an electrical installation are called the **extraneous conductive parts**

True ❑   False ❑

10 Structural steelwork, metal service pipes and heating radiators are called **exposed conductive parts**

True ❑   False ❑

Multiple Choice Assessment Questions

Tick the correct answer. Note that more than ONE answer may be correct

11 **In the SI system of units, the units of voltage current and resistance are:**
   a   volts, watts and newtons   ❑
   b   metre, kilogram and second   ❑
   c   volts, amps and ohms   ❑
   d   newton, joule and watt   ❑

12 **In the SI system of units, the units of length, mass and time are:**
   a   volts, watts and newtons   ❑
   b   metre, kilogram and second   ❑
   c   volts, amps and ohms   ❑
   d   newton, joule and watt   ❑

13 **In the SI system of units, the units of force, energy and power are:**
   a   volts, watts and newtons   ❑
   b   metre, kilogram and second   ❑
   c   volts, amps and ohms   ❑
   d   newton, joule and watt   ❑

14 **Electricity is generated in Power Stations at 25 kV. In the SI system of units 25 kV may be written as:**
   a   25 volts or 25 thousand volts   ❑
   b 25×10^{-3} amps or $25 \div 1{,}000$ amps   ❑
   c 25×10^{3} volts or 25,000 volts   ❑
   d   25 amps or 25 thousand amps   ❑

15 **Electronic equipment uses very small amounts of current. In the SI system of units twenty-five milliamperes may be written as:**
a   25 volts or 25 thousand volts ☐
b 25×10^{-3} amps or $25 \div 1,000$ amps ☐
c 25×10^{3} volts or 25,000 volts ☐
d   25 amps or 25 thousand amps ☐

16 **An insulator is a material in which the electrons are:**
a   very large compared with the nucleus ☐
b   positively charged to the nucleus ☐
c   tightly bound to the nucleus ☐
d   loosely bound to the nucleus ☐

17 **A conductor is a material in which the electrons are:**
a   very large compared with the nucleus ☐
b   positively charged to the nucleus ☐
c   tightly bound to the nucleus ☐
d   loosely bound to the nucleus ☐

18 **A 'good conductor' material has:**
a   a negative nucleus in the atoms of the material ☐
b   positive electrons available for current flow ☐
c   free electrons available for current flow ☐
d   no free electrons ☐

19 **A 'good insulator' material has:**
a   a negative nucleus in the atoms of the material ☐
b   positive electrons available for current flow ☐

c   free electrons available for current flow   ☐
d   no free electrons   ☐

20 The following materials are good conductors:
a   copper, perspex and glass   ☐
b   copper, brass and wood   ☐
c   copper, brass and aluminium   ☐
d   PVC, rubber and porcelain   ☐

21 The following materials are good insulators:
a   PVC, copper and aluminium   ☐
b   PVC, rubber and brass   ☐
c   PVC, rubber and porcelain   ☐
d   copper, gold and silver   ☐

22 An electric current in a circuit may also be described as a:
a   flow of atoms   ☐
b   difference of potential   ☐
c   resistance in the circuit   ☐
d   flow of free electrons   ☐

23 Most electrical cables are constructed in three parts, the:
a   conductor, copper and aluminium   ☐
b   conductor, insulation and flexible cord   ☐
c   conductor, insulation and sheath   ☐
d   conductor, outer sheath and protection   ☐

24 PVC insulated and sheathed cables and cords would be suitable for the following situations:
a   the fixed wiring in domestic installations   ☐
b   the fixed wiring in industrial installations   ☐

c flexible cords connecting domestic
appliances to a 13 A socket outlet   ❑

d an underground cable to a remote
building such as a domestic garage   ❑

**25 PVC/SWA cables would be suitable for the
following situations:**

a the fixed wiring in domestic installations   ❑

b the fixed wiring in industrial installations   ❑

c flexible cords connecting domestic
appliances to a 13 A socket outlet   ❑

d an underground cable to a remote
building such as a domestic garage   ❑

**26 When an electric current flows in an
electric circuit it can have one or more of
the following three effects:**

a voltage resistance and current   ❑

b steaming, smoking and getting hot   ❑

c heating, magnetic and chemical   ❑

d conduction, convection and radiation   ❑

**27 Using Ohm's Law, calculate the resistance
of a circuit in which the voltage was 230 V
and the current 5 A:**

a 21.7 ohm   ❑

b 46.0 ohm   ❑

c 460 ohm   ❑

d 1,150 ohm   ❑

**28 Using Ohm's Law, calculate the current
flowing in a 230 V kettle element of
resistance 19.166 ohm:**

a 8.33 A   ❑

b 12.00 A   ❑

c 16.66 A ☐

d 4408 A ☐

29 **Using Ohm's Law, calculate the voltage connected across a resistor of 1,000 ohm when a current of 3 milliamperes flows:**

a 3 mV ☐

b 3 V ☐

c 30 V ☐

d 300 V ☐

30 **Calculate the resistance of a one metre bar of silver, 1.5 mm² in cross-sectional area if the resistivity of silver is 16.4 × 10 (m):**

a 10.93×10^{-3} ohm ☐

b 10.93 milli-ohm ☐

c 91.46×10^{-3} ohm ☐

d 91.46 milli-ohm ☐

31 **Calculate the resistance of a one metre bar of iron, 1.5 mm² in cross-sectional area, if the resistivity of iron is 100 × 10 (m):**

a 15.00×10^{-3} ohm ☐

b 66.66×10^{-3} ohm ☐

c 66.66 milli-ohm ☐

d 15.00 milli-ohm ☐

32 **The resistance of the iron bar in Question 21 above, compared with the resistance of the silver bar in Question 20 is:**

a the iron bar has about 6 times less resistance ☐

b   the iron bar has about 6 times more
     resistance   ☐

c   the iron bar has about 15 times less
     resistance   ☐

d   the iron bar has about 15 times more
     resistance   ☐

**33 Two 6 ohm resistors are connected
first in series and then in parallel.
For each connection calculate the total
resistance:**

a   series 2 ohm   parallel 3 ohm   ☐
b   series 3 ohm   parallel 2 ohm   ☐
c   series 3 ohm   parallel 12 ohm   ☐
d   series 12 ohm   parallel 3 ohm   ☐

**34 Three resistors of 24, 40 and 60 ohms
are connected first in series and then in
parallel. For each connection calculate
the total resistance:**

a   series 124 ohm parallel 2.4 ohm   ☐
b   series 124 ohm parallel 12 ohm   ☐
c   series 124 ohm parallel 15 ohm   ☐
d   series 124 ohm parallel 124 ohm   ☐

**35 Three 2 ohm resistors are connected
first in series and then in parallel
across a 12 volt battery supply.
Calculate the current flowing for
each connection:**

a   series 2 A   parallel 18.18 A   ☐
b   series 8 A   parallel 12.5 A   ☐
c   series 12.5 A   parallel 8.0 A   ☐
d   series 18.18 A   parallel 2.0 A   ☐

36 **To correctly measure the current and voltage in a circuit, the meters must be connected to the load in the following way:**
a ammeter in series, voltmeter across the load ❑
b ammeter across the load, voltmeter in series ❑
c ammeter in series, voltmeter in parallel ❑
d ammeter in parallel, voltmeter in series ❑

37 **Magnetic energy causes:**
a like poles to attract ❑
b unlike poles to repel ❑
c like poles to repel ❑
d unlike poles to attract ❑

38 **A measure of the amount of material in a substance is called its:**
a force ❑
b gravity ❑
c mass ❑
d weight ❑

39 **The force which acts toward the centre of the earth is called:**
a force ❑
b gravity ❑
c mass ❑
d weight ❑

40 **Calculate the work done (WD) by a 50 kg bag of cement when it falls 10 metres from a scaffold to the ground. What do you think**

might be the consequences of this action?
What might be the consequences if the bag
fell on to a worker below? Assume the
acceleration due to gravity to be 9.81 m/s:

a   WD 3.996 kJ – bag remains intact ☐
b   WD 4.90 kJ – bag bursts ☐
c   WD 4.90 kJ – bag remains intact ☐
d   WD 50.96 kJ – bag bursts open ☐

The worker below would certainly be injured,
possibly seriously.

41 **Calculate the efficiency of a 1 kW electric
motor which takes 1200 W from the source
of supply:**

a   10.9% ☐
b   12.43% ☐
c   83.33% ☐
d   120% ☐

42 **Increasing the number of secondary turns
on a transformer connected to an A.C.
supply will:**

a   decrease the input voltage ☐
b   decrease the output voltage ☐
c   increase the input voltage ☐
d   increase the output voltage ☐

43 **The iron core of a transformer is:**

a   solid so as to increase the core
    magnetic flux ☐
b   laminated so as to increase the core
    magnetic flux ☐
c   solid in order to reduce the losses ☐
d   laminated in order to reduce the losses ☐

44 **The metal parts of a building structure are called:**

a   earthing   ☐
b   equipotential bonding   ☐
c   exposed conductive parts   ☐
d   extraneous conductive parts   ☐

45 **The metal parts of an electrical installation not normally live are called:**

a   earthing   ☐
b   equipotential bonding   ☐
c   exposed conductive parts   ☐
d   extraneous conductive parts   ☐

46 **The act of connecting exposed conductive parts to the earthing terminal of an installation is called:**

a   earthing   ☐
b   equipotential bonding   ☐
c   exposed conductive parts   ☐
d   extraneous conductive parts   ☐

47 **The process which maintains a potential of zero volts, between all exposed and extraneous parts is called:**

a   earthing   ☐
b   equipotential bonding   ☐
c   exposed conductive parts   ☐
d   extraneous conductive parts   ☐

48 **Pliers, cutters, a knife and a range of screwdrivers are the tools required in the electrotechnical industry for:**

a   erecting conduit   ☐
b   assembling tray   ☐

c stripping and connecting conductors ❑

d terminating an MI cable ❑

49 When visually inspecting an electrical power tool before using it, you notice minor damage to the case and the coloured conductors showing at the junction with the plug top. This power tool, in this condition should:

a not be used by the company trainee ❑

b only be used if the PAT test label is 'in date' ❑

c only be used by a 'competent person' ❑

d not be used until inspected and tested by a 'competent person' ❑

50 The most common cause of accidents at work is:

a gloves, boots and hard hats ❑

b sprains, strains and trap pains ❑

c slips, trips and falls ❑

d hook, line and sinker ❑

Health and Safety Application and Electrical Principles

Chapter 3 covers the topics described in the third core unit of the City & Guilds 2330 Syllabus for the Level 2 Certificate in Electrotechnical Technology

This Chapter describes safe systems of working and the principle of operation of some electrical machines, equipment and systems.

Health and Safety Applications

Avoiding Accidents in the Workplace

The Health & Safety at Work Act 1974 places a statutory and common law obligation on employers to take reasonable care of the health and safety of their workers. The Management of Health & Safety at Work Regulations 1999 places an obligation on employers to carry out "risk assessments" and, where necessary, to take action to eliminate or control risks. The Workplace (Health, Safety and Welfare) Regulations 1992 and the Construction Health, Safety and Welfare Regulations 1996 cover all aspects of the workplace and construction sites respectively. They include the requirement that all areas where people could fall from a height of two metres or above, are properly guarded. The latest HSE Regulations "Working at Height" were introduced in April 2005. The aim of these Regulations is to avoid working at height, if possible, but where this cannot be avoided, to use the best practicable means of ensuring the safety of those working at height. However, despite all the legislation, we know from the HSE statistics that accidents still occur in the workplace.

The most common causes of accidents in the workplace are:

◆   slips, trips and falls

◆   manual handling, that is, moving objects by hand

Figure 3.1   Slips, trips and falls are the most common causes of accidents in the workplace

◆   using equipment, machinery or tools

◆   storage of goods and materials which then become unstable and fall on someone

◆   fire

◆   electricity

◆   mechanical handling

To control the risk of an accident we usually:

◆   eliminate the cause, that means, do not do the job or procedure in an unsafe way

◆ substitute a procedure or product with less risk, that means finding a safer way to complete the job or procedure

◆ enclose the dangerous situation, that means fitting guards or screening off an area and only allowing trained and competent people into a potentially dangerous area

◆ put guards around a hazard, for example, placing guards in front of cutting and grinding wheels

◆ use safe systems of work, that means establishing written procedures for work that is potentially dangerous. These written procedures are sometimes called 'permits to work'

◆ supervise, train and give information to staff which leads to a 'competent' workforce

◆ if a hazard cannot be removed or minimised, then the employer must provide PPE. However, providing personal protective equipment to staff must be a last resort when the hazard cannot be removed in any other way. The PPE must be provided at the employer's expense

A **Hazard** is something with the **potential to cause harm**; for example, electric tools, working above ground level, wet or uneven floors, rotating parts.

A **Risk** is the **possibility of harm actually being done**. Is it a high, low or medium risk? Who is at risk, the office staff, electricians, the public? Is the risk adequately controlled?

A **positive, personal attitude to safety** reduces accidents at work. Always work and act responsibly and safely to protect yourself and others. Be aware of the hazards around you, the protection available to you and the means of preventing accidents.

Risk Assessment, the Process

We have already said that an employer must carry out risk assessments as a part of a robust Health and Safety policy. The HSE recommends five steps to any risk assessment.

Step 1

Look at what might reasonably be expected to cause harm. Ignore the trivial and concentrate only on significant hazards that could result in serious harm or injury. For example:

- Slipping, tripping or falling hazards, e.g. from poorly maintained or partly installed floors and stairs

- Fire, e.g. from flammable materials you might be using such as solvents

- Rotating parts of hand tools, e.g. drills

- Accidental discharge of cartridge operated tools

- Manual handling, e.g. lifting, moving or supporting loads

Step 2

Decide who might be harmed, do not list individuals by name. Just think about groups of people

HAZARD RISK ASSESSMENT	FLASH-BANG ELECTRICAL CO.
For Company name or site: _____ Address: _____ _____	Assessment undertaken by: _____ Signed: _____ Date: _____

STEP 5   Assessment review date: _____

STEP 1 List the hazards here	STEP 2 Decide who might be harmed

STEP 3 Evaluate (what is) the risk – is it adequately controlled? State risk level as low, medium or high	STEP 4 Further action – what else is required to control any risk identified as medium or high?

Figure 3.2   Hazard Risk Assessment – standard form

doing similar work or who might be affected by your work:

◆ Office staff

◆ Electricians

◆ Maintenance personnel

◆ Other contractors on site

Step 3

Evaluate what is the risk arising from an identified hazard. Is it adequately controlled or should more be done? Is the risk low, medium or high. Only low risk will be acceptable when the HSE Inspector comes to inspect your company records. Do the precautions already taken:

◆ meet the legal standards required

◆ comply with recognised industrial practice

◆ represent good practice

◆ reduce the risk as far as is reasonably practicable

If you can answer 'yes' to the above points then the risks are adequately controlled, but you need to state the pre-cautions that have been put in place, e.g. electric shock hazard from using portable equipment is reduced to low by PAT testing all equipment every 6 months.

Step 4

Further action – what more could be done to reduce those risks which were found to be inadequately controlled?

TRY THIS

To help you to be more aware of the hazards around you at work you might like to carry out a risk assessment on a situation you are familiar with at work, using the Standard form of Fig. 3.2 or your employer's standard form. Make a few photocopies and ask your Supervisor to help you, perhaps one lunch time.

Any hazard identified by a risk assessment as high risk must be brought to the attention of the person responsible for health and safety within the company.

Step 5

The assessment must be reviewed from time to time by the person responsible for health and safety.

Safe Manual Handling

There have been so many injuries over the years as a result of lifting, transporting or supporting loads by hand or bodily force that the Health & Safety Executive has introduced new legislation, the Manual Handling Operations Regulations 1992. These state that:

- if a job involves considerable manual handling, workers must be trained in the correct lifting procedure

- loads must not be lifted manually if it is more appropriate to use a mechanical aid

- always use a trolley, sack truck or wheelbarrow when these are available

- use good manual lifting techniques if the load must be lifted manually and avoid jerky movements

- only lift and carry what you can manage easily

- wear gloves to avoid rough or sharp edges

Good Manual Lifting Techniques

When manually lifting objects from the floor:

◆ bend at the hips and knees to get down to the object

◆ grasp the object firmly

◆ take account of its centre of gravity

◆ keep your back straight and head erect, use the powerful leg muscles to raise the object

◆ carry the load close to the body

Figure 3.3 Correct manual lifting and carrying procedure

Safe Working above Ground Level

Working above ground level is hazardous because there is a risk of falling. If the working platform is appropriate for the purpose, properly erected and in good condition, then the risk is low.

However, in 2004 the HSE statistics show that there were 67 fatal falls and almost 4,000 major injuries resulting from falls. They are the biggest single cause of workplace deaths and one of the main causes of major injury. If you fall from a height above two metres statistically you are 'very likely' to sustain a serious injury. To reduce accidents as a result of falls from height the HSE have introduced 'The Work at Height Regulations 2005'. They became law in April 2005.

The main hazards associated with working at height are people falling and objects falling on to people. The main aim of the Regulations is to:

◆ avoid working at height if possible

◆ no work must be done at height if it is safe and reasonably practicable to do it other than at height

◆ use work equipment to prevent falls where it is impossible to avoid working at height. That is guard rails and toe boards on scaffold platforms

◆ where the risk of a fall cannot be eliminated, robust platforms must be built so as to 'minimise' the distance and consequences should a fall occur. This may mean building a stepped pyramid type of platform at the work site when the risk is high

◆ risk assessments must be carried out

◆ everyone involved in work at height must be 'competent' and if being trained, must be supervised by a competent person

Ladders

The term ladder is generally taken to include stepladders and trestles. The use of ladders for working above ground level is only acceptable for access and work of a short duration. For work over an extended period, a temporary working platform or stage is inherently a much safer means of working above ground level.

There is extensive published guidance on the safe use of ladders which is summarised below:

◆ It is advisable to inspect any ladder before climbing it

◆ The ladder should be straight and firm without signs of any damage or cracks

◆ All rungs and tie rods must be in place

◆ Ladders must not be painted because the paint may hide any defects

This is important safety information

◆ Extension ladders must be erected in the closed position as shown in Fig. 3.4

Figure 3.4 Correct procedure for erecting long or extension ladders

◆ Each section of an extension ladder must overlap by at least two rungs

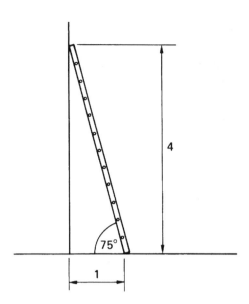

Figure 3.5 A correctly erected ladder

◆ The angle of the ladder to the building should be in the proportion 4 up to 1 out or 75° as shown in Fig. 3.5

◆ The top of the ladder must rest against a solid secure structure and not against a fragile or movable structure

◆ The top of the ladder must extend at least 1.05 m above the landing place or the highest rung on which the user has to stand

◆ Erect the ladder close to the work site and do not over-reach

◆ The ladder must stand on firm, level ground and be secured top and bottom

◆ All ladders should be tested and examined by a competent person at least yearly and the results recorded

Stepladders

The following precautions should be observed when using stepladders:

◆ They should be inspected before use

◆ Damaged, cracked or loose jointed stepladders should not be used

◆ They must be extended fully

◆ All four legs must rest firmly and squarely on firm ground

◆ The stepladder should be placed at right angles to the work wherever possible

◆ Do not stand on the top platform unless it is designed as a working platform

◆ do not use the top tread, tool shelf or rear part of the steps as a foot support

◆ Only one person should stand on the stepladder at any one time

◆ The stepladder must be suitable and of an appropriate grade for the intended use

Trestle Scaffold

Two pairs of trestles or 'A' frames spanned by scaffolding boards provide a simple working platform as shown in Fig. 3.6.

◆ As with stepladders, they must be erected on form level ground with the trestles fully opened

◆ The platform must be at least two boards or 450 mm wide

Figure 3.6 A trestle scaffold

◆ At least one third of the trestle must be above the working platform

◆ The scaffold boards must be of equal length and not overhang the trestles by more than four times their thickness

◆ The maximum span of the scaffold boards between the trestles depends upon the thickness of the boards. One metre for 32 mm boards, 1.5 m for 38 mm boards and 2.5 m for 50 mm boards

Mobile Scaffold Towers

Mobile scaffold towers are normally made from light aluminium tube, slotting sections together until the required height is reached. Mobile towers are fitted with four lockable wheels, static towers have flat plates instead of wheels. A mobile scaffold tower is shown in Fig. 3.7.

This is the preferred method of working above ground for extended periods. If accidents occur it is mainly as

Guard rail

Close boarded working platform

Toe boards

Diagonal bracing

Locking castors

Figure 3.7 A mobile scaffold tower

a result of poor standards of erection or misuse. Consider the following good practice:

◆ The person erecting the tower must be 'competent'

◆ Use the tower only on level, firm ground

◆ If the working platform is 2 m above ground it must be close boarded and fitted with guard rails and toe boards

◆ The taller the tower, the more likely it is to become unstable. Out-riggers can increase stability by effectively increasing the base area. Always keep within the manufacturer's safe working limits

◆ There must be a safe method of getting to and from the working platform. This is usually a built-in ladder which is climbed on inside the tower

◆ Wheel brakes must be on when the tower is in use

◆ Do not move the tower while it is occupied by people or there is material on the upper platform

◆ Push or pull the tower only from the base and look out for overhead obstructions

◆ Never extend the working platform with ladders or stepladders

◆ Ladders must not be leaned against the scaffold tower because this might push the tower over

◆ Ensure that the tower scaffold is regularly inspected and maintained by a trained and competent person

Safe Electrical Isolation and Lock Off

As an electrician working on electrical equipment you must always make sure that the equipment or circuit is electrically dead before commencing work to avoid receiving an electric shock and because:

◆ the Electricity at Work Regulations 1989 tell us that before work commences on electrical equipment it must be disconnected from the source of supply and that disconnection must be secure. A small padlock or the removal of the fuse or MCB will ensure the security of the disconnection

◆ the IEE Regulations (130-06-01) tell us that every circuit must be provided with a means of isolation

◆ larger pieces of equipment and electrical machines will often have an isolator switch close by which may be locked off

◆ to deter anyone from trying to re-connect the supply while work is being carried out, a sign 'Danger – Electrician at Work' should be displayed on the isolator or source of the supply in addition to the small padlock

◆ where a test instrument or voltage indicator such as that shown in Fig. 3.8 is used to prove the supply dead, the same device

Figure 3.8   Typical voltage indicator

Figure 3.9 Voltage proving unit

must be tested to prove it is still working by using a 'proving unit' such as that shown in Fig. 3.9

◆ the test leads and probes of the test instrument must comply with the Health & Safety Executive Guidance Note 38 giving adequate protection to the user as shown in Fig. 3.10

HBC fuses
and/or current limitation

Robust, flexible well-insulated leads

Barrier to prevent
access to live terminals

Minimum of
exposed metal

Shrouded or firmly
attached connectors

Figure 3.10 GS38 recommended test probes and Leads

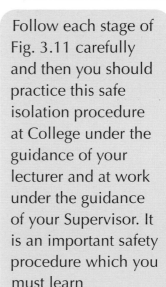

TRY
THIS

◆ a suitable safe electrical isolation procedure is shown in Fig. 3.11

Electrical Installation Principles

A.C. Theory

Commercial quantities of electricity for industry, commerce and domestic use are generated as A.C. in large Power Stations and distributed around the UK on the National Grid to the end user. D.C. electricity has many applications where portability or an emergency stand-by supply is important but for large quantities of power it has to be an A.C. supply.

Rotating a simple loop of wire or coils of wire between the poles of a magnet such as that shown

Follow each stage of Fig. 3.11 carefully and then you should practice this safe isolation procedure at College under the guidance of your lecturer and at work under the guidance of your Supervisor. It is an important safety procedure which you must learn

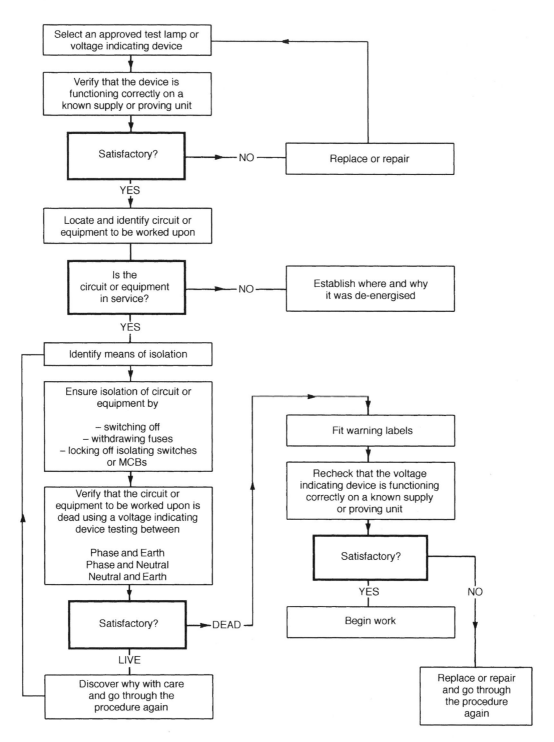

Figure 3.11 Safe electrical isolation procedure

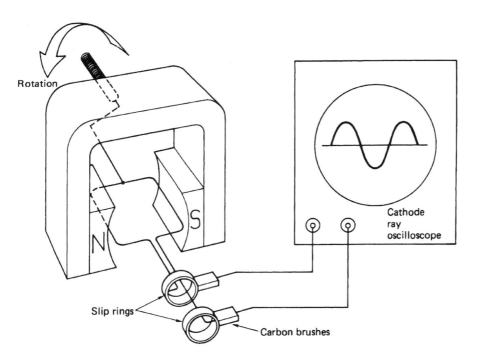

Figure 3.12   Simple A.C. generator or alternator

simplified in Fig. 3.12 will cut the north south lines of magnetic flux and induce an A.C. voltage in the loop or coils of wire as shown by the display on a cathode ray oscilloscope.

This is an A.C. supply, an alternating current supply. The basic principle of the A.C. supply generated in a Power Station is exactly the same as Fig. 3.12 except that powerful electromagnets are used and the power for rotation comes from a steam turbine.

Let us now look at some of the terms used in A.C. theory.

Phasor diagram

A phasor diagram or phasor is a straight line, having definite length and direction which represents to scale the voltage and current in an A.C. circuit.

You should memorise the A.C. theory which I will cover in the next section

Resistance

In any circuit, resistance is the opposition to current flow. Figure 3.13 shows the voltage and current relationships in resistive, inductive and capacitive circuits. Look at the left side of Fig. 3.13 which shows a resistor connected to an A.C. supply. You can see that when the voltage waveform reaches its maximum, so does the current. This always happens when resistive components are connected to an A.C. supply and we say that the voltage and current are "in phase" because they are always together. This is represented as a phasor diagram by the bottom left-hand sketch. You might like to think of the phasors as the minute and hour hands of a clock with rotation anti-clockwise. In this case the phasors are together showing that V and I are in phase.

Figure 3.13 Voltage and current relationship in resistive, inductive and capacitive circuits

Water heaters, electric fires and filament lamps are resistive circuits.

Inductance

Any coil of wire possesses inductance and so we say that the opposition to current flow in an inductive circuit is called '***inductive reactance***' symbol X_L measured in ohms.

When a current flows in a coil, it sets up its own voltage around the conductor which opposes the applied voltage. This causes the current to fall behind or 'lag' the applied voltage. You can see this in Fig. 3.13. Time is measured from left to right and so the current reaches its maximum value later than the voltage waveform. In fact, 90° later and so we say that in an inductive circuit, the current lags the voltage by 90°. This is represented on the phasor diagram as shown by the bottom centre sketch.

Inductive circuits are those which contain windings or coils such as electric motors, transformers or the choke of a discharge luminaire.

Capacitance

A capacitor is a component which stores an electric charge if a voltage is applied across it and so we say that the opposition to current flow in a capacitor circuit is called '***capacitive reactance***' symbol X_C measured in ohms.

When a capacitor is connected to the A.C. supply, it is continuously storing charge and then discharging as the supply moves through its positive to negative cycle. This causes the current to spring forward or to 'lead' the applied voltage. You can see this effect in Fig. 3.13. In fact, the current leads the applied voltage by 90°, just the opposite effect to an inductive circuit which is why this is important in A.C. circuit theory.

Capacitors are usually constructed from long strips of metal foil, like baking foil, insulated and then rolled up into a cylinder. You can see them in fluorescent fittings, discharge luminaires and sometimes fixed to an electric motor. The leading effect of the capacitor can be used to neutralise the lagging effect of inductors.

Power Factor

Power factor or pf is defined as the cosine of phase angle between the current and voltage. If the current lags the voltage as can be seen in the inductive circuit of Fig. 3.13 we say that the pf is lagging and if the current leads the voltage we say the pf is leading. The ideal situation is when the pf is neither lagging nor leading but is in phase. In this situation the pf is equal to 1.

To correct or put right the bad (lagging) power factor of an inductive circuit such as an electric motor or fluorescent light fitting, we would connect a capacitor (having a leading power factor) across the load. The leading pf of the capacitor neutralises the lagging pf of the inductive circuit bringing the overall pf of the circuit up to, or nearly up to 1. This is called power factor correction and 0.9, 0.95 or 1 are all acceptable values for a commercial, industrial or domestic supply.

Figure 3.14(a) shows the phasor diagram of an industrial load with a bad power factor. If a capacitor is connected in parallel with the load, the capacitor current I_c will lead the voltage by 90°. When the capacitor current is added to the load current as shown in Fig. 3.14(b) the resultant load current has a much improved power factor. Using a slightly bigger capacitor, the load current could be pushed up until it was 'in phase' with the voltage as shown in Fig. 3.14(c).

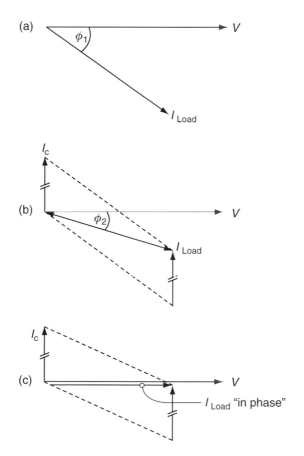

Figure 3.14   Power factor improvement using capacitors

Self and Mutual Inductance

If a coil of wire is wound on to an iron core as shown in Fig. 3.15 a magnetic field will become established

Figure 3.15   An inductive coil or choke

in the core when a current flows in the coil due to the switch being closed.

When the switch is opened, the current stops flowing and, therefore, the magnetic flux collapses. The collapsing magnetic flux induces an emf into the coil and this voltage appears across the switch contacts.

If you switch off a circuit containing fluorescent light fittings you can sometimes hear the discharge across the switch contacts (each fluorescent fitting contains a choke). The effect is known as *self-inductance*, or just *inductance* and is the property of any coil.

When two separate coils are placed close together – as they are in a transformer – a current in one coil produces a magnetic flux which links with the second coil. This induces a voltage in the second coil, and is the basic principle of the transformer action which is described later in this Chapter. The two coils in this case are said to possess *mutual inductance*, as shown in Fig. 3.16.

The emf induced in a coil such as that shown on the right-hand side in Fig. 3.16 is dependent upon the rate

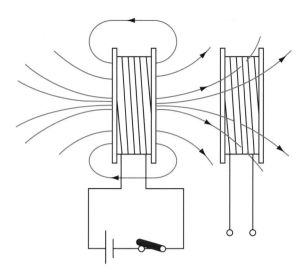

Figure 3.16   Mutual inductance between two coils

of change of magnetic flux and the number of turns on the coil. This principle finds an application in electric motors and transformers which we will discuss next.

Electrical Machines – Basic Operating Principles

Fluorescent Luminaires

A luminaire is equipment which supports an electric lamp and distributes or filters the light created by the lamp. It is essentially the "light fitting".

A lamp is a device for converting electrical energy into light energy. There are many types of lamps. General lighting service (GLS) lamps and tungsten halogen lamps use a very hot wire filament to create the light and so they also become very hot in use. Fluorescent tubes operate on the "discharge" principle; that is, the excitation of a gas within a glass tube. They are cooler in operation and very efficient in converting electricity into light. They form the basic principle of most energy efficient lamps.

A fluorescent lamp is a linear arc tube, internally coated with a fluorescent powder, containing a light low pressure mercury vapour and argon gas. The lamp construction is shown in Fig. 3.17.

Tube filled with argon and mercury vapour

Cathode coated with electron emitting material and fitted with cathode shield

Bi-pin cap

Glass, internally coated with fluorescent phosphor, cut away to reveal cathode

The arc radiates much more UV than visible light: almost all the light from a fluorescent tube comes from the phosphors

Figure 3.17

Passing a current through the electrodes of the tube produces a cloud of electrons that ionise the mercury vapour and the argon in the tube, producing invisible ultraviolet light and some blue light. The fluorescent powder on the inside of the glass tube is very sensitive to ultraviolet rays and converts this radiation into visible light.

Fluorescent luminaires require a simple electrical circuit to initiate the ionisation of the gas in the tube

Figure 3.18

and a device to control the current once the arc is struck and the lamp is illuminated. Such a circuit is shown in Fig. 3.18.

A typical application for a fluorescent luminaire is in suspended ceiling lighting modules used in many commercial buildings.

The Electrical Relay

A relay is an electromagnetic switch operated by a solenoid. We looked at the action of a solenoid in the last Chapter at Fig. 2.12. The solenoid in a relay operates a number of switch contacts as it moves under the electromagnetic forces. Relays can be used to switch circuits on or off at a distance remotely. The energising circuit, the solenoid, is completely separate to the switch contacts and, therefore, the relay can switch high voltage, high power circuits, from a low voltage switching circuit. This gives the relay many applications in motor control circuits, electronics and instrumentation systems. Figure 3.19 shows a simple relay.

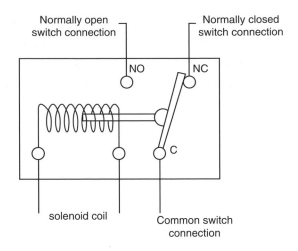

Figure 3.19 A Simple Relay

D.C. Motors

All electric motors work on the basic principle that when a current carrying conductor is place in a magnetic field it will experience a force. An electric motor uses this magnetic force to turn the shaft of the electric motor. Let us try to understand this action. Figure 3.20(a) shows the magnetic field set up around a current carrying conductor shown in cross-section. Figure 3.20(b) shows the

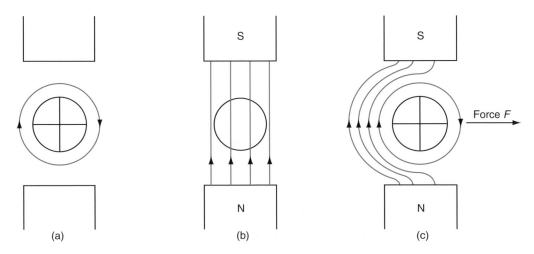

Figure 3.20 Force on a conductor in a magnetic field

magnetic field due to a permanent magnet in which is placed the conductor carrying no current. Figure 3.20(c) shows the effect of the combined magnetic fields which have become distorted and, because lines of magnetic flux behave like stretched elastic bands, a force F is exerted on the conductor. This is the force which turns the shaft on the electric motor.

A D.C. motor has a field winding wound on to the body or yoke of the motor and an armature winding which rotates and turns the motor shaft. Feeding the current into the armature, so that the magnetic field can be established, is the commutator and carbon brushes as shown in Fig. 3.21. D.C. motors are classified by the

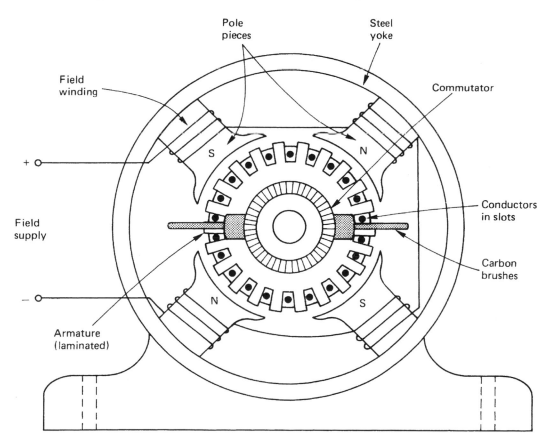

Figure 3.21   Showing D.C. machine construction

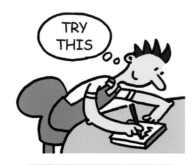

The next time you are using a 110 V electric drill to drill a wall, switch off the hammer action and listen to the sound of the drill. As you put pressure on the drill the speed will reduce because you are loading up the motor. Reduce the pressure and the drill will speed up because it is a series motor

way in which the field and armature windings are connected. Figure 3.22 shows the connections for a series motor. Because the windings are in series, a D.C. motor will also work satisfactorily on an A.C. supply. Small D.C. series motors are also called universal motors and are used for vacuum cleaners and hand drills because they have a high starting torque for a small motor, but if the load is increased, the speed reduces.

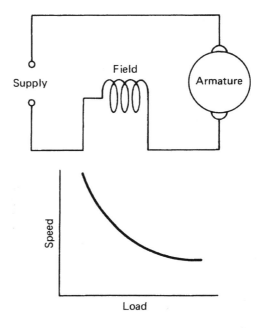

Figure 3.22   Series motor connections and characteristics

A.C. Motors

A.C. motors are also called induction motors because of their basic principle of operation. The A.C. supply is connected to the stator windings of the motor. These are the stationary windings of the motor, like the field winding of a D.C. motor. The A.C. supply sets up a rotating magnetic field in the stator, which causes the rotor to turn. Figure 3.23 shows the magnetic flux in

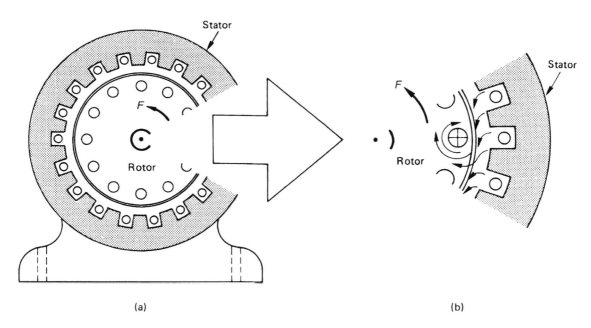

Stator

Rotor

Stator

F

Rotor

(a)

(b)

Figure 3.23 Segment taken out of induction motor

the stator and rotor creating the turning force or torque to drive the motor.

A D.C. motor always has a commutator and carbon brushes to connect the supply to the rotating part of the motor. These require maintenance and repair. An A.C. motor has no such equipment because the current is "induced" into the rotor by magnetic induction. A principle discovered by Michael Faraday. No carbon brushes or commutator is a great advantage in an A.C. machine and also the construction of the rotor makes an A.C. machine very robust.

Larger motors used in industry are connected to a three phase A.C. supply, while smaller motors are connected to a single phase A.C. supply.

A.C. motors have a relatively low starting torque and are used for constant speed applications from industrial motors to air extraction fans, fan heaters, central heating pumps, refrigerators and washing machines. Very small A.C. motors of less than 50 watts can be found in most domestic and business machines where single phase supplies are available. Figure 3.24 shows the construction of a small A.C. motor.

Figure 3.24   Shaded pole A.C. motor

Transformers

A transformer is an electrical machine which is used to change the value of an alternating voltage. They vary in size from miniature units used in electronics to huge power transformers used in power stations. A transformer will only work when an alternating voltage is connected. It will not normally work from a D.C. supply such as a battery.

A transformer, as shown in Fig. 3.25 consists of two coils, called the primary and secondary coils, or windings, which are insulated from each other and wound on to the same steel or iron core.

An alternating voltage applied to the primary winding produces an alternating current, which sets up an

Figure 3.25 A simple transformer

alternating magnetic flux throughout the core. This magnetic flux induces an emf in the secondary winding by mutual inductance, which was described earlier in this Chapter under the sub-heading 'Self and mutual inductance'. Since both windings are linked by the same magnetic flux, the induced emf per turn will be the same for both windings. Therefore, the emf in both windings is proportional to the number of turns. In symbols:

FOLLOW THIS MATHS

$$\frac{V_p}{N_p} = \frac{V_s}{N_s}$$

Where V_p = the primary voltage
  V_s = the secondary voltage
  N_p = the number of primary turns
  N_s = the number of secondary turns

moving the terms around we have a general expression for a transformer:

$$\frac{V_p}{V_s} = \frac{N_p}{N_s}$$

Using the general equation for a transformer given above, follow this maths carefully, step by step, in the following example.

EXAMPLE

A 230 V to 12 V emergency lighting transformer is constructed with 800 turns. Calculate the number of secondary turns required. Collecting the information given in the question into a usable form, we have:

$$V_p = 230\,V$$

$$V_s = 12\,V$$

$$N_p = 800$$

From the general equation:

$$\frac{V_p}{V_s} = \frac{N_p}{N_s}$$

the equation for the secondary turn is

$$N_s = \frac{N_p V_s}{V_p}$$

$$\therefore N_s = \frac{800 \times 12\,V}{230\,V} = 42 \text{ turns}$$

42 turns are required on the secondary winding of this transformer to give a secondary voltage of 12 V.

Types of Transformer

Step down Transformers are used to reduce the output voltage, often for safety reasons. Figure 3.26 shows a Step down transformer where the primary winding has twice as many turns as the secondary winding. The turns ratio is 2:1 and, therefore, the secondary voltage is halved.

Figure 3.26 A step down transformer

Step up Transformers are used to increase the output voltage. The electricity generated in a power station is stepped up for distribution on the National Grid Network. Figure 3.27 shows a Step up transformer where the primary winding has only half the number of turns as the secondary winding. The turns ratio is 1:2 and, therefore, the secondary voltage is doubled.

Figure 3.27 A step up transformer

Instrument Transformers are used in industry and commerce so that large currents and voltage can be measured by small electrical instruments.

A Current Transformer (or CT) has the large load currents connected to the primary winding of the transformer and the ammeter connected to the secondary winding. The ammeter is calibrated to take account of the turns ratio of the transformer, so that the ammeter displays the actual current being taken by the load when the ammeter is actually only taking a small proportion of the load current.

A Voltage Transformer (or VT) has the main supply voltage connected to the primary winding of the transformer and the voltmeter connected to the secondary winding. The voltmeter is calibrated to take account of the turns ratio of the transformer, so that the voltmeter displays the actual supply voltage.

Separated extra-low voltage (SELV) Transformers
If the primary winding and the secondary winding of a double wound transformer have a separate connection to earth, then the output of the transformer is effectively isolated from the input since the only connection between the primary and secondary windings is the magnetic flux in the transformer core. Such a transformer would give a very safe electrical supply which might be suitable for bathroom equipment such as shaver sockets and construction site 110 V tools, providing that all other considerations are satisfied, such as water ingress, humidity, IP protection and robust construction.

Generation, Transmission and Distribution of Electricity

Generation

Figure 3.12 earlier in this Chapter shows a simple A.C. generator or alternator producing an A.C. waveform. We generate electricity in large modern power stations using the same basic principle of operation. However, in place of a single loop of wire, the power station alternator has a three phase winding and powerful electromagnets. The prime mover is not, of course, a simple crank handle, but a steam turbine. Hot water is heated until it becomes superheated steam, which drives the vanes of a steam turbine which is connected to the alternator. The heat required to produce the steam may come from burning coal or oil or from a nuclear reactor. Whatever the primary source of energy is, it is only being used to drive a turbine which is connected to an alternator, to generate electricity.

Transmission

Electricity is generated in the power station alternator at 25 kV. This electrical energy is fed into a transformer to be stepped up to a very high voltage for transmission on the National Grid Network at 400 kV, 275 kV or 132 kV. These very high voltages are necessary because, for a given power, the current is greatly reduced, which means smaller grid conductors and the transmission losses are reduced.

The National Grid Network consists of over 5,000 miles of overhead aluminium conductors suspended from steel pylons which link together all the power stations. Figure 3.28 shows a transmission line steel pylon.

Figure 3.28   Transmission line steel pylon

Electricity is taken from the National Grid by appropriately located sub-stations which eventually transform the voltage down to 11 kV at a local sub-station. At the local sub-station the neutral conductor is formed for single phase domestic supplies and three phase supplies to shops, offices and garages. These supplies are usually underground radial supplies from the local sub-station but in rural areas we still see transformers and overhead lines suspended on wooden poles. Figure 3.29 gives an overview of the system from power station to consumer.

Distribution to the Consumer

The electricity leaves the local sub-station and arrives at the consumer's mains intake position. The final connections are usually by simple underground radial feeders

Figure 3.29   Simplified diagram of the distribution of electricity from power station to consumer

at 400 V/230 V. The 400 V/230 V is derived from the 11 kV/400 V sub-station transformer by connecting the secondary winding in star as shown in Fig. 3.30. The star point is earthed to an earth electrode sunk into the ground below the sub-station and from this point is taken the fourth conductor, the neutral. Loads connected between phases are fed at 400 V and those fed between one phase and neutral at 230 V. A three phase

Figure 3.30 Three phase four wire distribution

400 V supply is used for supplying small industrial and commercial loads such as garages, schools and blocks of flats. A single phase 230 V supply is usually provided for individual domestic consumers.

At the mains intake position the supplier will provide a sealed HBC fuse and a sealed energy meter to measure the consumer's electricity consumption. It is after this point that we reach the consumer's installation.

Balancing single phase loads

A three phase load such as a motor has equally balanced phases since the resistance of each phase winding will be the same. Therefore, the current taken by each phase will be equal. When connecting single phase loads to a three phase supply, care should be taken to distribute the single phase loads equally across the three phases so that each phase carries approximately the same current. Equally distributing the single phase loads across the three phase supply is known as 'balancing' the load. A lighting load of 18 luminaires would be 'balanced' if six luminaires were connected to each of the three phases.

Protecting Electrical Equipment, Circuits and People

Figure 3.31 The Provision of a Safe electrical installation, etc.

The provision of a safe electrical system is fundamental to the whole concept of using electricity in and around buildings safely. The electrical installation as a whole must be protected against overload and short circuit damage and the people using the installation must be protected against electric shock. An installation which meets the requirements of the IEE Regulations, Requirements for Electrical Installations, will be so protected. The method most universally used in the UK to provide for the safe use of electrical energy is earthed equipotential bonding coupled with automatic disconnection of the supply by fuses or MCBs. So let us look at these two essential safety elements.

You should learn these definitions and understand them.

Earthing and Bonding

Chapter 54 of the IEE Regulations describes the earthing arrangements for an electrical installation. Let us define some of the terms used in earthing and bonding.

Earth: The general mass of the planet earth is considered to be a large conductor at zero potential (potential means voltage in this case). The act of earthing connects together all metalwork, other than that intended to carry current, to the general mass of earth so that a dangerous potential difference cannot exist between different metal parts, or between metal parts and earth.

Earthing: The IEE Regulations define earthing as the act of connecting the exposed conductive parts of an installation to the main earthing terminal of the installation.

Exposed Conductive Parts: The IEE Regulations define these as a conductive part which may be touched and which is not live under normal conditions, but may become live under fault conditions. This means the metalwork of an appliance or the metalwork of the electrical installation such as the conduit, trunking or metal boxes of the electrical system, all of which must be connected to the main earthing terminal of the installation.

Circuit Protective Conductor (CPC): This is a protective conductor connecting exposed conductive parts to the main earthing terminal. It will be a green and yellow insulated conductor of appropriate size.

Extraneous Conductive Parts: This is the structural steelwork of a building and other service pipes used for gas, water, etc. (radiators and sinks). They do not form a part of the electrical installation, but may introduce a potential to the electrical installation. To eliminate this hazard we provide equipotential bonding.

Equipotential Bonding: This is an electrical connection which maintains exposed conductive parts and extraneous conductive parts at the same potential. To do this, we connect a green and yellow insulated cable of appropriate size to all extraneous parts and connect this to the main earthing terminal of the installation. By connecting to earth all metalwork not intended to carry current, a safe path is provided for any leakage currents which can be detected and disconnected by fuses, circuit breakers and RCDs.

A good earth path, that is a low resistance earth path, will allow high fault currents to flow, which will cause protective devices to operate quickly and remove the potential hazard quickly.

Overcurrent Protection

All circuit conductors must be protected against overcurrent, that is, a current exceeding the rated value (Regulation 431-01-01). Fuses and circuit breakers provide overcurrent protection when situated in the live conductor. They must not be connected in the neutral conductor.

Overcurrent conditions arise because of an overload or a short circuit in the electrical circuit.

By definition an overload current occurs in a circuit which is carrying more current than it was designed to carry. The excess current may be a result of too many pieces of equipment being connected to the circuit or because a piece of equipment has become faulty. An overload current will result in currents of two or three times the rated current flowing. This will cause the cable temperature to rise, leading to an increased risk of fire.

By definition a short circuit current occurs in a circuit as a result of a fault or damage to the circuit which could not have been predicted before the event. The short circuit current may be the result of a nail being driven through an energised cable, making contact with the live conductor and either the neutral or earth conductors. A short circuit current will result in currents hundreds of times greater than the rated current flowing. To avoid the risk of fire or electric shock, these overcurrents must be interrupted quickly and the circuit made dead.

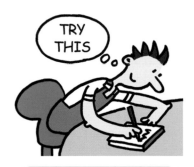

Try writing out these definitions – just the essential bits to help you to remember them – for overcurrent, overload, short circuit current and a fuse.

Devices which provide overcurrent protection are:

◆ Semi-enclosed fuses to BS 3036

◆ Cartridge fuses to BS 1361

◆ MCBs (miniature circuit breakers) to BS 3871

By definition a fuse is the weakest link in the circuit. Under fault conditions it will melt when an overcurrent flows, protecting the circuit conductors from damage.

Semi-enclosed Fuses (BS 3036)

The semi-enclosed fuse consists of a fuse wire, called the fuse element, secured between two screw terminals in a fuse carrier. The fuse element is connected in series with the load and the thickness of the element is sufficient to carry the normal rated circuit current. When a fault occurs an overcurrent flows and the fuse element becomes hot and melts or 'blows'.

This type of fuse is illustrated in Fig. 3.32. The fuse element should consist of a single strand of plain or

Figure 3.32   A semi-enclosed fuse

tinned copper wire having a diameter appropriate to the current rating of the fuse. This type of fuse was very popular in domestic installations, but less so these days because of their disadvantages.

Advantage of semi-enclosed fuses:

◆ They are very cheap compared with other protective devices both to install and to replace

◆ There are no mechanical moving parts

◆ It is easy to identify a 'blown' fuse

Disadvantages of semi-enclosed fuses:

◆ The fuse element may be replaced with wire of the wrong size either deliberately or by accident

◆ The fuse element weakens with age due to oxidisation, which may result in a failure under normal operating conditions

◆ The circuit cannot be restored quickly since the fuse element requires screw fixing

◆ They have low breaking capacity since, in the event of a severe fault, the fault current may vaporise the fuse element and continue to flow in the form of an arc across the fuse terminals

◆ They are not guaranteed to operate until up to twice the rated current is flowing

◆ There is a danger from scattering hot metal if the fuse carrier is inserted into the base when the circuit is faulty

Cartridge Fuses (BS 1361)

The cartridge fuse breaks a faulty circuit in the same way as a semi-enclosed fuse, but its construction eliminates some of the disadvantages experienced with an open-fuse element. The fuse element is encased in a glass or ceramic tube and secured to end-caps which are firmly attached to the body of the fuse so that they do not blow off when the fuse operates. Cartridge fuse construction is illustrated in Fig. 3.33. With larger size cartridge fuses, lugs or tags are sometimes brazed on the end-caps to fix the fuse cartridge mechanically to the carrier. They may also be filled with quartz sand to absorb and extinguish the energy of the arc when the cartridge is brought into operation.

Figure 3.33   Cartridge fuse

Advantages of Cartridge Fuses:

◆   They have no mechanical moving parts

◆   The declared rating is accurate

◆   The element does not weaken with age

◆   They have small physical size and no external arcing which permits their use in plug tops and small fuse carriers

◆ Their operation is more rapid than semi-enclosed fuses. Operating time is inversely proportional to the fault current

◆ They are easy to replace

Disadvantages of Cartridge Fuses:

◆ They are more expensive to replace than fuse elements that can be re-wired

◆ They can be replaced with an incorrect cartridge

◆ The cartridge may be shorted out by wire or silver foil in extreme cases of bad practice

◆ It is not possible to see if the fuse element is broken

Miniature Circuit Breakers (BS 3871)

The disadvantage of all fuses is that when they have operated they must be replaced. An MCB overcomes this problem since it is an automatic switch which opens in the event of an excessive current flowing in the circuit and can be closed when the circuit returns to normal.

An MCB of the type shown in Fig. 3.34 incorporates a thermal and magnetic tripping device. The load current flows through the thermal and the electromagnetic devices in normal operation but under overcurrent conditions they activate and trip the MCB.

The circuit can be restored when the fault is removed by pressing the ON toggle. This latches the various mechanisms within the MCB and 'makes' the switch

Figure 3.34   MCBs – B Breaker, fits Wylex Standard consumer unit, Courtesy of Wylex.

contact. The toggle switch can also be used to disconnect the circuit for maintenance or isolation or to test the MCB for satisfactory operation.

Advantages of MCBs:

◆   They have factory set operating characteristics

◆   Tripping characteristics and therefore circuit protection is set by the installer

◆   The circuit protection is difficult to interfere with

◆   The circuit is provided with discrimination

◆   A faulty circuit may be quickly identified

◆   A faulty circuit may be easily and quickly restored

If you have read and understood the whole of this Chapter, you have completed all of the underpinning knowledge requirements of the third core unit in the City & Guilds 2330 Syllabus for the Level 2 Certificate in Electrotechnical Technology.

When you have completed the practical assessments required by the City & Guilds Syllabus, which you are probably doing at your local College, you will be ready to tackle the on-line assessments. So, to prepare you for the On-Line Assessment, try the following Assessment Questions.

◆ The supply may be safely restored by an unskilled operator

Disadvantages of MCBs:

◆ They are expensive

◆ They contain mechanical moving parts and therefore require regular testing to ensure satisfactory operation under fault conditions

Assessment Questions

Identify the statements as true or false. If only part of the statement is false, tick false

1 The most common cause of accidents in the workplace are:
   ◆ Slips, trips and falls
   ◆ Moving objects by hand
   ◆ Storing equipment badly, which then becomes unstable
   ◆ An electric shock
   True ❏   False ❏

2 A hazard is defined as the possibility of harm actually being done. It may be high or low but to the Health and Safety Inspector, only low will be acceptable.
   True ❏   False ❏

3 Risk is something that might cause harm to a worker. For example, wet floors, working above ground or using machinery
   True ❏   False ❏

4 Many accidents are caused by lifting and moving heavy objects manually (that is, by hand). If a job involves a lot of manual handling, workers must be:
   ◆ Trained to move loads safely
   ◆ Or be provided with a suitable mechanical aid such as a sack truck
   True ❏   False ❏

5 When working above ground, ladders must only be used for access to the workplace or for working above ground for short periods only.
True ❑   False ❑

6 When erecting a ladder, it must be made secure and have an angle against the wall of 90° or 1 up to every 4 out.
True ❑   False ❑

7 A safe and secure electrical isolation procedure will always include a means of 'locking off' the electrical supply.
True ❑   False ❑

8 When connecting single phase loads across a three phase supply they should be equally balanced so that the currents in the three phase supply remain approximately the same.
True ❑   False ❑

9 By definition, a fuse is the weakest link in a circuit. Under fault conditions it will melt, protecting the circuit from damage.
True ❑   False ❑

10 By definition, an overload current occurs as a result of a fault, and a short circuit current occurs in a circuit which is carrying more current than it was designed to carry.
True ❑   False ❑

Multiple Choice Assessment Questions

Tick the correct answer. Note that more than ONE answer may be correct

11 Health and Safety legislation requires all employers who employ more than five workers to:
- a   pay each worker 10% above the National rate for the job ☐
- b   display a Health and Safety Law poster ☐
- c   prepare a written Health and Safety policy statement ☐
- d   carry out risk assessments ☐

12 Slips, trips and falls:
- a   do not happen at work because of safety legislation ☐
- b   always happen to someone else ☐
- c   are one of the most common causes of accidents in the workplace ☐
- d   must be reported to the HSE if they result in an absence from work of more than three days ☐

13 A 'competent' worker is one who:
- a   cannot do the job or task ☐
- b   can do the job or task more quickly than anyone else ☐
- c   has been trained to do a job or task successfully ☐
- d   is quarrelsome and likely to cause an argument at work ☐

14 Hazard may be defined as:
a   anything that can cause harm ☐
b   the chance, large or small, of harm
actually being done ☐
c   someone who has the necessary
training and expertise to safely carry
out an activity ☐
d   the rules and regulations of the working
environment ☐

15 Risk may be defined as:
a   anything that can cause harm ☐
b   the chance, large or small, of harm
actually being done ☐
c   someone who has the necessary
training and expertise to safely carry
out an activity ☐
d   the rules and regulations of the
working environment ☐

16 Hazard Risk Assessment is:
a   the harm which might be done to an
employee not wearing PPE ☐
b   the hazard created when someone lifts
very heavy object ☐
c   the process of systematically examining
the workplace for possible dangers ☐
d   the risk of harm being done to someone
in the workplace ☐

17 A positive attitude to safety at work:
a   is the duty of every employer ☐
b   is the duty of every employee ☐
c   increases accidents at work ☐
d   reduces accidents at work ☐

18 **Manual handling is the process of:**
   a following instructions from a reference
      book ❏
   b following instructions from manufacturers'
      data sheets ❏
   c lifting, transporting or supporting loads
      by hand or bodily force ❏
   d moving a heavy load on a sack truck or
      other mechanical aid ❏

19 **The min hazards associated with working
   at height are:**
   a ladders not being secured top and bottom ❏
   b extension ladders not being fully extended ❏
   c people falling ❏
   d objects falling on to people ❏

20 **The angle of a ladder to the building upon
   which it is resting should be in the
   proportions of:**
   a 1 up to 4 out ❏
   b 4 up to 75 out ❏
   c 4 up to 1 out ❏
   d 75 up to 4 out ❏

21 **The angle which a correctly erected ladder
   should make with level ground is:**
   a 41° ❏
   b 45° ❏
   c 57° ❏
   d 75° ❏

22 **Extension ladders should be erected:**
   a one section at a times ❏
   b in the closed position ❏

c   in the open position ☐

d   against a solid structure ☐

23 **Ladders must extend above the landing place or highest rung on which the user will stand by:**

a   1.00 m ☐

b   1.05 m ☐

c   4.00 m ☐

d   75.00 m ☐

24 **All ladders, including step ladders, must be tested and inspected at least yearly by:**

a   the managing director of the company ☐

b   the site supervisor or engineer ☐

c   any competent person ☐

d   the senior electrical trainee ☐

25 **To use stepladders safely:**

a   all four legs must rest firmly and squarely on firm ground ☐

b   always stand on the top platform ☐

c   always stand on the tool platform ☐

d   they must be fully extended ☐

26 **The preferred method of working above ground level for an extended period is:**

a   a stepladder ☐

b   an extension ladder ☐

c   a trestle scaffold ☐

d   a scaffold tower ☐

27 **For good stability mobile towers must have a base width to tower height ratio of:**

a   1:2 ☐

b   1:3 ☐

c 1:4 ❑

d 1:5 ❑

28 **To verify or prove a successful electrical isolation you would use a:**

a voltage indicator such as that shown in Fig. 3.8 ❑

b voltage proving unit such as that shown in Fig. 3.9 ❑

c set of GS 38 test leads ❑

d small padlock ❑

29 **To secure an electrical isolation you would use a:**

a voltage indicator such as that shown in Fig. 3.8 ❑

b voltage proving unit such as that shown in Fig. 3.9 ❑

c set of GS 38 test leads ❑

d small padlock ❑

30 **Where a test instrument or voltage indicator is used to prove a supply dead, the same device must be tested to show that it still works using a:**

a voltage indicator such as that shown in Fig. 3.8 ❑

b voltage proving unit such as that shown in Fig. 3.9 ❑

c set of GS 38 test leads ❑

d small padlock ❑

31 **To give adequate protection to the person carrying out a safe isolation procedure, the test instrument must incorporate a:**

a voltage indicator such as that shown in Fig. 3.8 ❑

    b voltage proving unit such as that shown
        in Fig. 3.9 ☐
    c set of GS 38 test leads ☐
    d small padlock ☐

32 **When a resistor circuit is connected to an A.C. supply:**
    a an alternating current flows ☐
    b the current and voltage are 'in phase' ☐
    c the current falls behind or 'lags' the voltage ☐
    d the current springs forward or 'leads'
        the voltage ☐

33 **When an inductive coil is connected to an A.C. supply:**
    a an alternating current flows ☐
    b the current and voltage are 'in phase' ☐
    c the current falls behind or 'lags' the voltage ☐
    d the current springs forward or 'leads' the
        voltage ☐

34 **When a capacitor is connected to an A.C. supply:**
    a an alternating current flows ☐
    b the current and voltage are 'in phase' ☐
    c the current falls behind or 'lags' the voltage ☐
    d the current springs forward or 'leads' the
        voltage ☐

35 **The cosine of the phase angle between the current and voltage is one definition of:**
    a resistance R ☐
    b inductive reactance X_L ☐
    c capacitive reactance X_C ☐
    d power factor pf ☐

36 **The opposition to current flow in a coil connected to an A.C. supply is:**
   a   resistance R ☐
   b   inductive reactance X_L ☐
   c   capacitive reactance X_C ☐
   d   power factor pf ☐

37 **The opposition to current flow in a capacitive A.C. circuit is:**
   a   resistance R ☐
   b   inductive reactance X_L ☐
   c   capacitive reactance X_C ☐
   d   power factor pf ☐

38 **Power factor correction is applied to an A.C. circuit in order to create the conditions where:**
   a   current leads the voltage ☐
   b   voltage leads the current ☐
   c   voltage and current are 'out of phase' ☐
   d   voltage and current are 'in phase' ☐

39 **Power factor correction for a fluorescent luminaire is achieved by connecting across the mains supply a:**
   a   choke ☐
   b   starter canister ☐
   c   ballast ☐
   d   capacitor ☐

40 **An electromagnetic switch operating a number of electrical contacts is one definition of:**
   a   A.C. motors ☐
   b   D.C. machines ☐

c   a relay ☐

d   a transformer ☐

41 **When a current carrying conductor is placed in a magnetic field it will experience a force. This is the basic principle of:**

a   A.C. motors ☐

b   D.C. machines ☐

c   a relay ☐

d   a transformer ☐

42 **Connecting an A.C. supply to the stator windings of the machine induces a rotating magnetic field which causes the rotor to turn. This is the basic principle of:**

a   A.C. motors ☐

b   D.C. machines ☐

c   a relay ☐

d   a transformer ☐

43 **An electrical machine has two separate windings on a common iron core. An A.C. voltage in one winding induces an A.C. voltage in the other winding, which is proportional to the number of turns. This is the basic principle of:**

a   A.C. motors ☐

b   D.C. machines ☐

c   a relay ☐

d   a transformer ☐

44 **A step down transformer has a turns ratio of 800 to 42. When a 230 V supply is connected the secondary voltage will be:**

a   5.47 V ☐

b   12.00 V ☐

c   19.00 V   ☐
d   43.83 V   ☐

45 **A step up transformer has a turns ratio of 1:10. If the primary voltage is 100 V the secondary voltage will be:**
a   1 V   ☐
b   10 V   ☐
c   100 V   ☐
d   1000 V   ☐

46 **A step down transformer has a turns ratio of 10:1. If the primary voltage is 100 V, the secondary voltage will be:**
a   1 V   ☐
b   10 V   ☐
c   100 V   ☐
d   1000 V   ☐

47 **A construction site transformer has a turns ratio of 535:256. When connected to a 230 V supply the transformer will deliver a secondary voltage of:**
a   2.09 V   ☐
b   110.06 V   ☐
c   480.60 V   ☐
d   595.50 V   ☐

48 **An isolating transformer for a tungsten halogen dichroic reflector lamp has a turns ratio of 38:2. Calculate the secondary voltage when connected to a 230 V supply:**
a   12.1 V   ☐
b   19.0 V   ☐
c   43.7 V   ☐
d   76.0 V   ☐

49 **A voltage transformer is connected to the 230 V mains supply so that it might supply an electrical measuring instrument operating at 2 V. The turns ratio of this transformer will therefore be:**

a   1:46   ❏

b   46:1   ❏

c   1:115   ❏

d   115:1   ❏

50 **An isolating transformer can be found:**

a   in local sub-stations operating at
    11 kV : 415 V   ❏

b   in large commercial power stations   ❏

c   in sub-stations connecting power lines to
    the National Grid   ❏

d   in a bathroom shaver unit   ❏

51 **When connecting single phase loads to a three phase supply, we must take care to distribute the single phase loads equally across the three phases so that each phase carries approximately the same current. This is called:**

a   generation of the phase loads   ❏

b   transmission of the phases   ❏

c   distribution of the load   ❏

d   balancing of the load   ❏

52 **The metal structural steelwork of a building is called:**

a   the general mass of earth   ❏

b   the circuit protective conductor (CPC)   ❏

c   exposed conductive parts   ❏

d   extraneous conductive parts   ❏

53 **The protective conductor connecting exposed conductive parts of equipment to the main earthing terminal is called:**
a   the general mass of earth ☐
b   the circuit protective conductor (CPC) ☐
c   exposed conductive parts ☐
d   extraneous conductive parts ☐

54 **The trunking and conduit of an electrical installation are called:**
a   the general mass of earth ☐
b   the circuit protective conductor (CPC) ☐
c   exposed conductive parts ☐
d   extraneous conductive parts ☐

55 **The metalwork of a piece of electrical equipment is called:**
a   the general mass of earth ☐
b   the circuit protective conductor (CPC) ☐
c   exposed conductive parts ☐
d   extraneous conductive parts ☐

56 **An electrical connection which maintains extraneous conductive parts at the same potential is called:**
a   CPC (circuit protective conductor) ☐
b   earth conductors ☐
c   equipotential bonding ☐
d   supplementary bonding ☐

57 **An overload current may be defined as:**
a   a current in excess of at least 15 A ☐
b   a current which exceeds the rated value
    in an otherwise healthy circuit ☐

   c an overcurrent resulting from a fault
between live and neutral conductors ☐

   d a current in excess of 60 A ☐

58 A short circuit may be defined as:

   a a current in excess of at least 15 A ☐

   b a current which exceeds the rated value
in an otherwise healthy circuit ☐

   c an overcurrent resulting from a fault
between live and neutral conductors ☐

   d a current in excess of 60 A ☐

**59 It is the weakest link in the circuit. Under
fault conditions it will melt, protecting the
circuit conductors from damage. This is
one description of:**

   a an electromagnetic relay ☐

   b an MCB (miniature circuit breaker) ☐

   c a fuse ☐

   d an isolating switch ☐

**60 It is an automatic switch which opens
when an overcurrent flows in the circuit.
This is one description of:**

   a a fuse ☐

   b an MCB ☐

   c an isolating switch ☐

   d a thermostat ☐

CHAPTER
4

Installation (Building and Structures)

Chapter 4 covers the topics described in Occupational Unit 4 Installation (Buildings and Structures) of the City & Guilds 2330 Syllabus for the Level 2 Certificate in Electrotechnical Technology

This Chapter is concerned with the underlying principles related to electrical installation work. Understanding the laws and regulations, the different types of installation and cabling and equipment used in electrical installation work.

Regulations and Responsibilities

In Chapter 1 of this Book we looked at a number of the regulations which control the electrotechnical industries. The Electricity at Work (EAW) Regulations are legally binding Regulations which concern all aspects of electrical systems, equipment and installations, which have been or are to be energised.

Electricity at Work Regulations and Codes of Practice

The EAW Regulations came into force on the 1st April 1990. The purpose of the Regulations is to require precautions to be taken against the risk of death or personal injury from electricity in work activities.

The EAW Regulations are made under the Health & Safety at Work Act 1974 and are statutory regulations (see Chapter 1 for a description of Statutory Laws). In the introduction at Section 7 of the EAW Regulations it sets out the position of the IEE Regulations in the following terms:

> The Institution of Electrical Engineers Regulations for Electrical Installations (the IEE Wiring Regulations) are non-statutory regulations relating

principally to the design, selection, erection and inspection and testing of electrical installations. The IEE Wiring Regulations is a code of practice which is widely recognised and accepted in the UK and compliance with them is likely to achieve compliance with relevant aspects of the EAW Regulations 1989.

If a contract specifies that the work will be carried out in accordance with BS 7671, the IEE Wiring Regulations, then this would be legally binding and the IEE Wiring Regulations will then become a legal requirement of the contract.

Figure 4.1 Electrical Regulations are enforced by Law

IEE Regulations (BS 7671)

The first edition of the IEE Regulations was issued in 1882 as the Rules and Regulations for the Prevention of Fire Risks Arising from Electric Lighting. In the intervening 125 years there have been many new editions and we are currently using the 16th Edition.

The main reason for incorporating the IEE Wiring Regulations into British Standard 7671 was to create harmonisation with European Standards. The IEE Wiring Regulations *On Site Guide* (BS 7671: 2001) are compliant with European Standards. British Standards having a BS EN number refers to a European harmonised standard and all such standards will become common throughout Europe.

The IEE Wiring Regulations (BS 7671) is the Electricians' Bible and provides the authoritative framework for anyone working in the electrotechnical industry.

To assist workers in the electrotechnical industry with their understanding of the relevant regulations many guidance booklets have been published, particularly:

◆ The *On Site Guide* published by the IEE

◆ Guidance Note 1: Selection and erection of equipment

◆ Guidance Note 2: Isolation and Switching

◆ Guidance Note 3: Inspection and Testing

◆ Guidance Note 4: Protection against Fire

◆ Guidance Note 5: Protection against Electric Shock

◆ Guidance Note 6: Protection against Overcurrent

◆ Guidance Note 7: Special Locations

All the above publications are published by the IEE and are available from IEE Publications, Michael Faraday House, Six Hills Way, Stevenage, SG1 2AY. Telephone (01438) 755540 or at www.iee.org/books

The 'Electrician's Guide to Good Electrical Practice', known as a 'toolbox guide' is published by the Trade Union Amicus.

On-Site Communications

Read through the "Communications and technical information" section of Chapter 1 before going on to this new work.

Good communication is about transferring information from one person to another. How many hours or days did you spend on a particular job last week? How does your boss know how many hours of work you put in on that job, so that a charge to the customer for your time can be made? How much material did you use on that job last week? How does your boss know how much material you used, so that a charge to the customer can be made for it?

Most electrical companies have standard forms which help them to keep track of time put in and materials used. When completing standard forms, follow the instructions given and make sure that your writing is legible – print if it makes your writing clearer. Finally, read through the form to make sure that you have completed all the relevant sections. Now, let us look at five standard forms used by most electrotechnical companies.

Time Sheets

A time sheet is a standard form completed by each employee to inform the employer of the actual time spent working on a particular contract or site. This helps the employer to bill the hours of work to an individual job. It is usually a weekly document and includes the number of hours worked, the name of the job and any travelling expenses claimed. Office personnel require time sheets such as that shown in Fig. 4.2 so that wages can be made up.

Job Sheets

A job sheet or job card such as that shown in Fig. 4.3 carries information about a job which needs to be done, usually a small job. It gives the name and address of the customer, contact telephone numbers, often a job reference number and a brief description of the work to be carried out. A typical job sheet work description might be:

◆ Job 1 – Upstairs lights not working

◆ Job 2 – Funny fishy smell from kettle socket in kitchen

The time spent on each job and the materials used are sometimes recorded on the job sheets, but alternatively, a daywork sheet can be used. This will depend upon what is normal practice for the particular electrical company. This information can then be used to 'bill' the customer for work carried out.

Day	Job number and/or Address	Start time	Finish time	Total hours	Travel time	Expenses

TIME SHEET

FLASH-BANG ELECTRICAL

Employee's name (Print) _____

Week ending _____

Day	Job number and/or Address	Start time	Finish time	Total hours	Travel time	Expenses
Monday						
Tuesday						
Wednesday						
Thursday						
Friday						
Saturday						
Sunday						

Employee's signature _____ Date _____

Figure 4.2 Time Sheet

```
┌─────────────────────────────────────────────┐
│  JOB SHEET              FLASH-BANG           │
│  Job Number _____ ELECTRICAL           │
├─────────────────────────────────────────────┤
│  Customer name _____ │
│                                              │
│  Address of job _____ │
│                                              │
│                 _____ │
│                                              │
│                 _____ │
│                                              │
│  Contact telephone no. _____ │
├─────────────────────────────────────────────┤
│  Work to be carried out _____ │
│                                              │
│                 _____ │
│                                              │
│                 _____ │
│                                              │
│                 _____ │
├─────────────────────────────────────────────┤
│   Any special instructions/conditions/materials used │
└─────────────────────────────────────────────┘
```

Figure 4.3 Job Sheets

Daywork Sheets

Daywork is one way of recording variations to a contract, that is, work done which is outside the scope of the original contract. If daywork is to be carried out, the site supervisor must first obtain a signature from the client's representative, for example, the Architect, to authorise the extra work. A careful record must then be kept on the daywork sheets of all extra time and materials used so that the client can be billed for the extra work and materials. A typical daywork sheet is shown in Fig. 4.4.

FLASH-BANG ELECTRICAL

DAYWORK SHEET

Client name _____

Job number/ref. _____

Date	Labour	Start time	Finish time	Total hours	Office use

Materials quantity	Description	Office use

Site supervisor or F.B. Electrical Representative responsible for carrying out work _____

Signature of person approving work and status e.g.

Client ☐   Architect ☐   Q.S. ☐   Main contractor ☐   Clerk of works ☐

Signature _____

Figure 4.4   Daywork Sheets

Delivery Notes

When materials are delivered to site, the person receiving the goods is required to sign the driver's 'Delivery Note'. This record is used to confirm that goods have been delivered by the supplier, who will then sent out an invoice requesting payment, usually at the end of the month.

The person receiving the goods must carefully check that all items stated on the Delivery Note have been delivered in good condition. Any missing or damaged items must be clearly indicated on the Delivery Note before signing because, by signing the Delivery Note the person signing is saying 'yes, these items were delivered to me as my company's representative on that date and in good condition and I am now responsible for those goods'. Suppliers will replace materials damaged in transit, provided that they are notified within a set period, usually three days. The person receiving the good should try to quickly determine their condition – has the packaging been damaged – does the container 'sound' like it might contain broken items? It is best to check at the time of delivery if possible or as soon as possible after delivery and within the notifiable period. Electrical goods delivered to site should be handled carefully and stored securely until they are installed. Copies of Delivery Notes should be sent to Head Office so that payment can be made for the goods received.

Reports

On large jobs, the foreman or supervisor is often required to keep a report of the relevant events which happen on the site; for example, how many people from the company you work for are working on site each day; what goods were delivered; whether there were any breakages or accidents and records of site meetings attended. Some firms have two separate documents, a site diary to record all daily events and a weekly report which is a summary of the week's events extracted from the site diary. The site diary remains on site and the weekly report is sent to Head Office to keep managers informed of the work's progress.

Electricity Supply Systems

The Electricity supplies to houses, shops, offices and small industrial consumers is nominally set at 230 V single phase and 400 V three phase. The nominal voltage must be maintained by the supplier within a tolerance range of plus or minus ten per cent (±10%). So, a domestic supply must be maintained by the supplier within 207 V and 253 V for single phase supplies and between 360 V and 440 V for three phase supplies. All EU countries agreed to change to these values from 2005. The frequency is maintained at 50 cycles per second over 24 hours so that electric clocks remain accurate.

The electricity supply to domestic, commercial and small industrial consumers is usually protected at the incoming service cable position by a 100 A HBC fuse. Other items of equipment at this position are the energy meter to record the electricity consumption and a consumer unit/fuseboard to provide the protection for the final circuits and the earthing arrangements for the installation. An efficient and effective

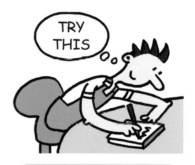

TRY THIS

To understand where electricity supplies come from, you should re-read the section 'Generation, Transmission and Distribution of Electricity' in Chapter 3. You should also look at Fig. 3.27 before starting on this section.

earthing system is essential to allow protective devices to operate quickly and effectively.

The IEE Regulations (BS 7671) gives details of the earthing arrangements in Section 542. Five systems are described, but only three electricity supply systems are suitable for public supplies and we will, therefore, only concern ourselves with these three supply systems.

Cable Sheath Earth Supplies (TN-S System)

This is one of the most common types of supply system to be found in the UK, where the electricity company's supply is provided by underground cables. The neutral and protective conductors are separate throughout the system. The protective earth conductor is the metal sheath and armour of the underground cable and this is connected to the consumer's main earthing terminal. All extraneous conductive parts of the installation, gas pipes, water pipes and any lightning protective system are connected to the protective conductor via the main earthing terminal of the installation. The arrangement is shown in Fig. 4.5.

Protective Multiple Earthing Supplies (TN-C-S system)

This type of underground supply is becoming increasingly popular to supply new installations in the UK. It is more commonly referred to as protective multiple earthing (PME). The supply cable uses a combined

Figure 4.5 Cable Sheath Earth Supplies (TN-S system) showing earthing and bonding arrangements

protective earth and neutral conductor. At the supply intake point a consumer's main earthing terminal is formed by the supply provider by connecting the earthing terminal to the neutral conductor. All extraneous conductive parts of the installation, gas pipes, water pipes and any lightning protective system are then connected to the main earthing terminals. Thus phase to earth faults are effectively converted into phase to neutral faults. The arrangement is shown in Fig. 4.6.

Figure 4.6 Protective Multiple Earthing (PME) Supplies (TN-C-S System) showing earthing and bonding arrangements

No Earth Provided Supplies (TT System)

This is the type of supply more often found when the installation is fed from overhead cables. The supply authorities do not provide an earth terminal and the installation's circuit protective conductors must be connected to earth via an earth electrode provided by the consumer. An effective earth connection is sometimes difficult to obtain and in most cases a residual current device is provided when this type of supply is used. The arrangement is shown in Fig. 4.7.

Figure 4.7 No Earth Provided Supplies (TT System) showing earthing and bonding arrangements

Wiring and Lighting Circuits

Table 1A in Appendix 1 of the IEE's *On Site Guide* deals with the assumed current demand of points and states that for lighting outlets we should assume a current equivalent to a minimum of 100 W per lampholder. This means that for a domestic lighting circuit rated at 5 A or 6 A a maximum of 11 or 12 lighting outlets could be connected to each circuit. In practice, it is usual to divide the fixed lighting outlets into two or more circuits of seven or eight outlets each. In this way the whole installation is not plunged into darkness if one lighting circuit fails.

Lighting circuits are usually wired in 1.0 mm or 1.5 mm cable using either a loop-in or joint-box method of installation. The loop-in method is universally employed with conduit installations or when access from above or below is prohibited after installation, as is the case with some industrial installations or blocks of flats. In this method the only joints are at the switches or lighting points, the live conductors being looped from switch to switch and the neutrals from one lighting point to another.

The use of junction boxes with fixed brass terminals is the method often adopted in domestic installations, since the joint boxes can be made accessible but are out of sight in the loft area or under floorboards.

All switches and ceiling roses must contain an earth connection (Regulation 471-09-02) and the live conductors must be broken at the switch position in order to comply with the polarity Regulations (713-09-01). A ceiling rose may only be connected to installations operating at 250 V maximum and must only accommodate one flexible cord unless it is specially designed to take more than one (553-04-02). Lampholders must comply with Regulation (553-03-02) and be suspended from flexible cords capable of suspending the mass of the luminaire fixed to the lamp-holder (554-01-01).

The type of circuit used will depend upon the installation conditions and the customer's requirements. One light controlled by one switch is called one-way switch control (see Fig. 4.8).

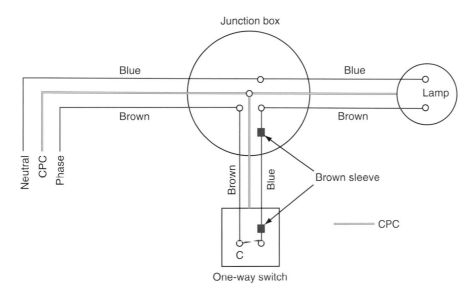

Figure 4.8   One-way switch control

A room with two access doors might benefit from a two-way switch control (see Fig. 4.9) so that the lights may be switched on or off at either position.

Fixing Positions of Switches and Sockets

Part M of the Building Regulations requires switches and socket outlets in dwellings to be installed so that all persons, including those whose reach is limited, can easily reach them. The recommendation is that

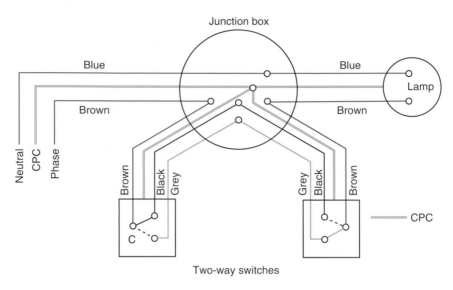

Figure 4.9   Two-way switch control

they should be installed in habitable rooms at a height of between 450 mm and 1200 mm from the finished floor level. This is shown in Fig. 4.10. The guidance given applies to all new dwellings but not to re-wires. However, these recommendations will undoubtedly 'influence' decisions taken when re-wiring dwellings.

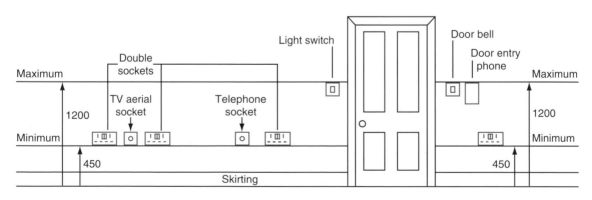

Figure 4.10   Fixing positions of switches and socket outlets

Socket Outlet Circuits

A plug top is connected to an appliance by a flexible cord which should normally be no longer than 2 m (Regulation 553-01-07). Pressing the plug top into a socket outlet connects the appliance to the source of supply. Socket outlets therefore provide an easy and convenient method of connecting portable electrical appliances to a source of supply.

Socket outlets can be obtained in 15, 13, 5 and 2 A ratings but the 13 A flat pin type complying with BS 1363 is the most popular for domestic installations in the United Kingdom. Each 13 A plug top contains a cartridge fuse to give maximum potential protection to the flexible cord and the appliance which it serves.

Socket outlets may be wired on a ring or radial circuit and, in order that every appliance can be fed from an adjacent and convenient socket outlet, the number of sockets is unlimited provided that the floor area covered by the circuit does not exceed that given in Table 8 A, Appendix 8 of the *On Site Guide* and Figs 4.11 and 4.12 of this book.

Radial Circuits

In a radial circuit each socket outlet is fed from the previous one. Live is connected to live, neutral and earth to earth at each socket outlet. The fuse and cable sizes are given in Table 8A of Appendix 8 of the *On Site Guide* but circuits may also be expressed with a block diagram as shown in Fig. 4.12. The number of permitted socket outlets is unlimited but each radial

Figure 4.11 Block diagram of radial socket circuits

Figure 4.12 Block diagram of ring socket circuits

circuit must not exceed the floor area stated and the known or estimated load.

Where two or more circuits are installed in the same premises, the socket outlets and permanently connected equipment should be reasonably shared out among the circuits so that the total load is balanced.

When designing ring or radial circuits, special consideration should be given to the loading in kitchens, which may require separate circuits. This is because the maximum demand of current-using equipment in kitchens may exceed the rating of the circuit cable and protection devices. Ring and radial circuits may be used for domestic or other premises where the demand of the current-using equipment is estimated not to exceed the rating of the protective devices for the chosen circuit.

Ring Circuits

Ring circuits are very similar to radial circuits in that each socket outlet is fed from the previous one, but in ring circuits the last socket is wired back to the source of supply. Each ring final circuit conductor must be looped into every socket outlet or joint box which forms the ring and must be electrically continuous throughout its length. **The number of permitted socket outlets is unlimited** but each ring circuit must not cover more than 100 m of floor area.

The circuit details are given in Table 8A, Appendix 8 of the *On Site Guide* but may also be expressed by the block diagram given in Fig. 4.12.

Socket Outlet Numbers

The Regulations allow us to install an unlimited number of socket outlets, the restriction being that each circuit must not exceed a given floor area as shown in Figs 4.11 and 4.12.

These days most households have lots of domestic appliances and electronic equipment, so how many sockets should be installed? Ultimately this is a matter

Figure 4.13 Electrician installing socket outlet circuits

for the customer and the electrical designer but most consumer organisations, the house builders NHBC and the Royal Society for the Prevention of Accidents (ROSPA) make the following general recommendations:

◆ The hard wiring for a single socket outlet is the same as the hard wiring for a double socket outlet. So, always install a double switched socket outlet unless there is a reason not to.

◆ Kitchens will require between six and ten double sockets, fitted both above and below the work surface for specific appliances

◆ Utility room – two double sockets

◆ Sittingrooms will require between six and ten double sockets with one double socket situated next to any telephone outlet to power telecommunication equipment and two double sockets adjacent to the TV aerial outlet for TV, video and DVD supplies

◆ Double bedrooms – four to six double sockets

◆ Single bedrooms – four to six double sockets

◆ Hallways – two double sockets with one situated next to any telephone outlet

◆ Home Office – six double sockets

◆ Garage – two double sockets

Cables and Enclosures

Power and lighting circuit conductors are contained within cables or enclosures. Part 5 of the IEE Regulations tells us that electrical equipment and materials must be chosen so that they are suitable for the installed conditions, taking into account temperature, the presence of water, corrosion, mechanical damage, vibration or exposure to solar radiation. Therefore, PVC insulated and sheathed cables are suitable for domestic installations but for a cable requiring mechanical protection and suitable for burying underground, a PVC/SWA cable would be preferable. These two types of cable are shown in Figs 2.2 and 2.3 in Chapter 2 of this book.

Mineral insulated (MI) cables are waterproof, heat-proof and corrosion resistant with some mechanical protection. These qualities often make it the only cable choice for hazardous or high temperature installations such as oil refineries, chemical works, boiler houses and petrol pump installations. An MI cable with terminating gland and seal is shown in Fig. 4.14.

Figure 4.14   MI Cable with terminating gland and seal

The FP 200 cable is another specialist cable. It is a fire resistant cable, primarily intended for use in fire alarm and emergency lighting installations. Its appearance is very similar to an MI cable in that it is constructed as a thin pencil size tube but the outer sheath is made from a robust thermoplastic material and is much easier to terminate than an MI cable.

We will look at wiring enclosures in the next section but first let us look at the new wiring colours for all fixed wiring which came into force on the 1st April 2006.

New Wiring Colours

On the 31st March 2004 the IEE published Amendment No. 2 to BS 7671: 2001 which specified new cable core colours for all fixed wiring in United Kingdom electrical installations. These new core colours will 'harmonise' the United Kingdom with the practice in mainland Europe.

Existing Fixed Cable Core Colours:

◆ *Single phase* – red phase conductors, black neutral conductors and green combined with yellow for earth conductors

◆ *Three phase* – red, yellow and blue phase conductors, black neutral conductors and green combined with yellow for earth conductors

These core colours must not be used after 31st March 2006

New (harmonised) Fixed Cable Core Colours:

◆ *Single phase* – brown phase conductors, blue neutral conductors and green combined with yellow for earth conductors (just like the existing flexible cords)

◆ *Three phase* – brown, black and grey phase conductors, blue neutral conductors and green combined with yellow for earth conductors

These core colours may be used from 31st March 2004

Extensions or alterations to existing *single phase* installations do not require marking at the interface between the old and new fixed wiring colours. However, a warning notice must be fixed at the consumer unit or distribution fuse board which states:

Caution – this installation has wiring colours to two versions of BS 7671. Great care should be taken before undertaking extensions, alterations or repair that all conductors are correctly identified.

Size of Conductor

Appendix 4 of the IEE Regulations (BS: 7671) and Appendix 6 of the IEE On Site Guide contain tables for determining the current carrying capacities of conductors. However, for standard domestic circuits, Table 4.1 gives a guide to cable size.

Table 4.1 Cable Size for Standard Domestic Circuits

Type of Final Circuit	Cable size (Twin and earth)	MCB rating, Type B (A)	Maximum floor area covered by circuit (m²)	Maximum length of cable run (m)
Fixed Lighting	1.0	6	–	40
Fixed Lighting	1.5	6	–	60
Immersion Heater	2.5	16	–	30
Storage Radiator	2.5	16	–	30
Cooker (oven only)	2.5	16	–	30
13 A Socket outlets (Radial circuit)	2.5	20	50	30
13 A Socket outlets (Ring circuit)	2.5	32	100	90
13 A Socket outlets (Radial circuit)	4.0	32	75	35
Cooker (oven and hob)	6.0	32	–	40
Shower (up to 7.5 kw)	6.0	32	–	40
Shower (up to 9.6 kw)	10	40	–	40

In this Table, I am assuming a standard 230 V domestic installation, having a sheathed earth or PME supply terminated in a 100 A HBC fuse at the mains position. Final circuits are fed from a consumer unit, having Type B, MCB protection and wired in PVC insulated and sheathed cables with copper conductors having a grey thermoplastic PVC outer sheath or a white thermosetting cable with LSF (low smoke and fume properties). I am also assuming that the surrounding temperature throughout the length of the circuit does not exceed 30°C and the cables are run singly and clipped to a surface.

Wiring Systems and Enclosures

The final choice of a wiring system must rest with those designing the installation and those ordering the work, but whatever system is employed, good workmanship and the use of proper materials is essential for compliance with the Regulations (IEE Regulation 130-02-01). The necessary skills can be acquired by an electrical trainee who has the corrected attitude and dedication to the craft.

PVC Insulated and Sheathed Cable Installations

PVC insulated and sheathed wiring systems are used extensively for lighting and socket installations in domestic dwellings. Mechanical damage to the cable caused by impact, abrasion, penetration, compression or tension must be minimised during installation (Regulation 522-06-01). The cables are generally fixed, using plastic clips incorporating a masonry nail, which means the cables can be fixed to wood, plaster or brick with almost equal ease. Cables should be run horizontally or vertically, not diagonally, down a wall. All links should be removed so that the cable is run straight and neatly between clips fixed at equal distances providing adequate support for the cable so that it does not become damaged by its own weight, as shown in Table 4.2. Where cables are bent, the radius of the bend should not cause the conductors to be damaged.

Table 4.2   Spacing of cable supports. Reproduced from the IEE *On Site Guide* by kind permission of the Institution of Electrical Engineers

Table 4A Spacings of supports for cables in accessible positions

Overall diameter of cable*	Maximum spacings of clips							
	Non-armoured thermosetting, thermoplastic or lead sheathed cables				Armoured cables		Mineral insulated copper sheathed or aluminium sheathed cables	
	Generally		In caravans					
	Horizontal[†]	Vertical[†]	Horizontal[†]	Vertical[†]	Horizontal[†]	Vertical[†]	Horizontal[†]	Vertical[†]
	2	3	4	5	6	7	8	9
mm	mm	mm	mm	mm	mm	mm	mm	mm
Not exceeding 9	250	400	250 (for all sizes)	400 (for all sizes)	–	–	600	800
Exceeding 9 and not exceeding 15	300	400			350	450	900	1200
Exceeding 15 and not exceeding 20	350	450			400	550	1500	2000
Exceeding 20 and not exceeding 40	400	550			450	600	–	–

Note: For the spacing of supports for cables having an overall diameter exceeding 40 mm, and for single-core cables having conductors of cross-sectional area 300 mm² and larger, the manufacturer's recommendations should be observed.

* For flat cables taken as the dimension of the major axis.

† The spacings stated for horizontal runs may be applied also to runs at an angle of more than 30 from the vertical. For runs at an angle of 30° or less from the vertical, the vertical spacings are applicable.

Terminations or joints in the cable may be made in ceiling roses, junction boxes or behind sockets or switches, provided that they are enclosed in a non-ignitable material, are properly insulated and are mechanically and electrically secure (IEE Regulation 526). All joints must be accessible for inspection and maintenance when the installation is completed.

Where PVC insulated and sheathed cables are concealed in walls, floors or partitions, they must be provided with a box incorporating an earth terminal at each outlet position. Figure 4.15 shows a typical concealed PVC sheathed wiring system.

Figure 4.15 A concealed PVC sheathed wiring system

To identify the most probable cable routes, Regulation 522-06-06 tells us that outside a zone formed by a 150 mm border all around a wall edge, cables can only be run horizontally or vertically to a point or accessory unless they are contained in a substantial earthed enclosure such as a conduit, which can withstand nail penetration, as shown in Fig. 4.16.

Where holes are drilled in floor joists to accommodate cable runs, they must meet the requirements shown in Fig. 4.17.

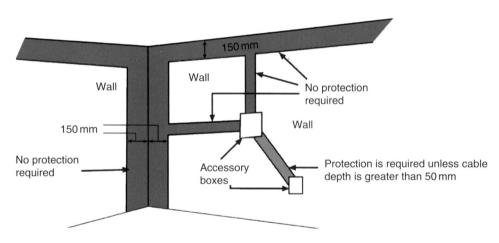

Figure 4.16 Permitted cable routes

Notes:

1. *Maximum diameter of hole should be 0.25 x joist depth.*
2. *Holes on centre line in a zone between 0.25 and 0.4 x span.*
3. *Maximum depth of notch should be 0.125 x joist depth.*
4. *Notches on top in a zone between 0.1 and 0.25 x span.*
5. *Holes in the same joist should be at least 3 diameters apart.*

Figure 4.17 Correct installation of cables in floor joists

Conduit Installations

A conduit is a tube, channel or pipe in which insulated conductors are contained. The conduit, in effect, replaces the PVC outer sheath of a cable, providing mechanical protection for the insulated conductors. A conduit installation can be re-wired easily or altered at any time, and this flexibility, coupled with mechanical protection, makes conduit installations popular for commercial and industrial applications.

There are three types of conduit used in electrical installation work: steel, PVC and flexible.

Steel Conduit

Steel conduit offers the conductors within a great deal of protection from mechanical damage. Steel conduit installations therefore, find an application in industrial environments.

PVC Conduit

PVC conduit used on typical electrical installations is heavy gauge standard impact tube manufactured to BS 4607. The conduit size and range of fittings are the same as those available for metal conduit. PVC conduit is most often joined by placing the end of the conduit into the appropriate fitting and fixing with a PVC solvent adhesive. PVC conduit can be bent by hand using a bending spring of the same diameter as the inside of the conduit.

The advantages of a PVC conduit system are that it can be installed much more quickly than steel conduit and is non-corrosive, but it does not have the mechanical strength of steel conduit. Since PVC conduit is an insulator it cannot be used as the CPC and a separate conductor must be run to every outlet. It is not suitable for installations subjected to temperatures below −5°C or above 60°C. Where luminaires are suspended from PVC conduit boxes, precautions must be taken to ensure that the lamp does not raise the box temperature or that the mass of the luminaire supported by each box does not exceed the maximum recommended by the manufacturer (IEE Regulation 522-01). PVC conduit also expands much more than metal conduit and so long runs require an expansion coupling to allow for conduit movement and help to prevent distortion during temperature changes.

All conduit installations must be erected first before any wiring is installed (IEE Regulation 522-08-02).

A limit must be placed on the number of bends between boxes in a conduit run and the number of cables which may be drawn into a conduit to prevent the cables being strained during wiring. Appendix 5 of the *On Site Guide* gives a guide to the cable capacities of conduits and trunking.

Flexible Conduit

Flexible conduit is made of interlinked metal spirals often covered with a PVC sleeving. The tubing must not be relied upon to provide a continuous earth path and, consequently, a separate CPC must be run either inside or outside the flexible tube (Regulation 543-02-01).

Flexible conduit is used for the final connection to motors so that the vibration of the motor are not transmitted throughout the electrical installation and to allow for modifications to be made to the final position and drive belt adjustments.

Trunking Installations

A trunking is an enclosure provided for the protection of cables which is normally square or rectangular in cross-section, having one removable side. Trunking may be thought of as a more accessible conduit system and for industrial and commercial installations it is replacing the larger conduit size. A trunking system can have great flexibility when used in conjunction with conduit; the trunking forms the background or framework for the installation, with conduits running from the trunking to the point controlling the current using apparatus.

Trunking is supplied in 3 m lengths and various cross-sections measured in millimetres from 50 × 50 up to 300 × 150. Most trunking is available in either steel or plastic.

Metallic Trunking

Metallic trunking is formed from mild steel sheet, coated with grey or silver enamel paint for internal use or a hot-dipped galvanised coating where damp conditions might be encountered. A wide range of accessories are available, such as 45° bends, 90° bends, tee and four-way junctions for speedy on-site assembly. Alternatively, bends may be fabricated in lengths of trunking, as shown in Fig. 4.18. This may be necessary or more convenient if a bend or set is non-standard, but it does take more time to fabricate bends than merely to bolt on standard accessories.

Figure 4.18 Alternative Trunking bends

Cable Tray Installations

Cable tray is a sheet-steel channel with multiple holes. The most common finish is hot-dipped galvanised but PVC coated tray is also available. It is used extensively on large industrial and commercial installations for supporting MI and SWA cables which are laid on the cable tray and secured with cable ties through the tray holes.

Cable tray should be adequately supported during installation by brackets which are appropriate for the particular installation. The tray should be bolted to the brackets with round-headed bolts and nuts, with the round head inside the tray so that cables drawn along the tray are not damaged.

The tray is supplied in standard widths from 50 mm to 900 mm and a wide range of bends, tees and reducers are available. Figure 4.19 shows a factory-made 90° bend at B. The tray can also be bent using a cable tray bending machine to create bends such as that shown at A in Fig. 4.19.

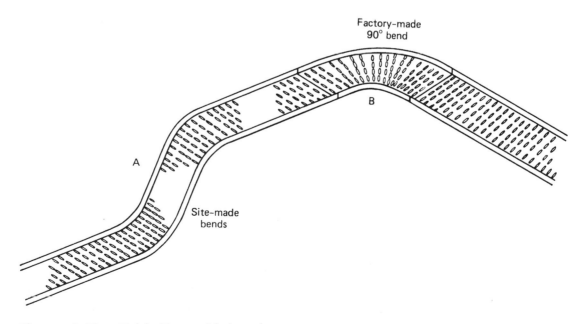

Figure 4.19   Cable Tray with bends

PVC/SWA Installations

A PVC/SWA cable is shown in Fig. 2.3.

MI Cable Installations

Mineral insulated cables are available with bare sheaths or with a PVC oversheath. The cable sheath provides sufficient mechanical protection for all but the most severe situations, where it may be necessary to fit a steel sheath or conduit over the cable to give extra protection, particularly near floor level in some industrial situations. (Figure 4.14 shows an MI cable termination.)

The cable may be laid directly in the ground, in ducts, on cable tray or clipped directly to a structure. It is not affected by water, oil or the cutting fluids used in engineering and can withstand very high temperatures or even fire. The cable diameter is small in relation to its current carrying capacity and it should last indefinitely if correctly installed because it is made from inorganic materials. These characteristics make the cable ideal for emergency circuits, boiler-houses, furnaces, petrol stations and chemical plant installations.

Special Installations

All electrical installations and installed equipment must be safe to use and free from the dangers of electric shock, but some installations require special consideration because of the inherent dangers of the installed conditions. The danger may arise because of the corrosive or explosive nature of the atmosphere, because the installation must be used in damp or low temperature conditions or because there is a need to provide additional mechanical protection for the electrical system. In this section we will consider some of the installations which require special consideration.

Bathroom Installations

In rooms containing a fixed bath tub or shower basin, additional regulations are specified. This is to reduce the risk of electric shock to people in circumstances where body resistance is lowered because of contact with water. The Regulations can be found in Section 601 and can be summarised as follows:

◆ Socket outlets must not be installed and no provision is made for connection of portable appliances

◆ Only shaver sockets which comply with BS EN 61184 or BS EN 60238, that is, those which contain an isolating transformer may be installed

◆ Every switch must be inaccessible to anyone using the bath or shower unless it is of the cord-operated type

◆ There are restrictions as to where appliances, switchgear and wiring accessories may be installed. See Zones for bath and shower-rooms below

◆ A supplementary bonding conductor must be provided in addition to the main equipotential bonding shown in Figs 4.5 to 4.7

Figure 4.20 Bathroom installations require special consideration

Zones for Bath and Shower Rooms

Locations that contain a bath or shower are divided in zones or separate areas as shown in Fig. 4.21.

Zone 0 – the bath tub or shower basin itself, which can contain water and is, therefore, the most dangerous zone

Zone 1 – the next most dangerous zone in which people stand in water

Zone 2 – the next most dangerous zone in which people might be in contact with water

Zone 3 – people are least likely to be in contact with water but are still in a potentially dangerous environment

Figure 4.21 Cross-section through bathroom showing zones

Electrical equipment and accessories are restricted within the zones:

Zone 0 – being the most potentially dangerous zone for all practical purposes, no electrical equipment can be installed in this zone. However, the Regulations permit that where SELV fixed equipment cannot be located elsewhere, it may be installed in this zone

Zone 1 – water heaters, showers and shower pumps and SELV fixed equipment

Zone 2 – luminaires, fans and heating appliances and equipment from Zone 1 plus shaver units to BS EN 60742

Zone 3 – fixed appliances are allowed plus the equipment from Zones 1 and 2

Outside Zones – appliances are allowed plus accessories except socket outlets

If under floor heating is installed in these areas it must have an overall earthed metallic grid or the heating cable must have an earthed metallic sheath which must be supplementary bonded.

Supplementary Bonding

Modern plumbing methods make considerable use of non-metals (PTFE tape on joints for example). There-fore, the metalwork of water and gas installations cannot be relied upon to be continuous throughout.

The IEE Regulations describe the need to consider additional or supplementary bonding in situations where there is a high risk of electric shock (for example, in kitchens and bathrooms). In rooms containing a fixed bath or shower, supplementary bonding con-ductors **must** be installed to reduce to a minimum the risk of an electric shock (Regulation 601-04-04). Bonding conductors in domestic premises will nor-mally be of 4 mm^2 copper with PVC insulation to com-ply with Regulations 547-03-01 to 03 and must be connected between all exposed metalwork (for exam-ple, between metal baths, bath and sink taps, shower fittings, metal waste pipes and radiators, as shown in Fig. 4.22.

The bonding connection must be made to a cleaned pipe, using a suitable bonding clip. Fixed at or near the connection must be a permanent label saying **'Safety electrical connection – do not remove'** (Regulation 514-13-01) as shown in Fig. 4.23.

Figure 4.22   Supplementary bonding in bathrooms to metal pipework

Figure 4.23   Typical earth bonding clamp

Temporary Installations (Construction Sites)

Figure 4.24   Construction Sites require Special Consideration

Temporary electrical supplies provided on construction sites can save many man-hours of labour by providing the energy required for fixed and portable tools and lighting which speeds up the completion of a project. However, construction sites are dangerous places and the temporary electrical supply which is installed to assist the construction process must comply with all of the relevant wiring regulations for permanent installations (Regulation 110-01-01). All equipment must be of a robust construction in order to fulfil the on-site electrical requirements while being exposed to rough handling, vehicular nudging, the wind, rain and sun.

All socket outlets, plugs and couplers must be of the industrial type to BS 4343 and specified by Regulation 604-12-02 as shown in Fig. 4.25.

(a)

(b)

Figure 4.25   100 V Distribution unit and cable connectors

IEE Regulation 604-02-02 recommends the following voltages for the distribution of electrical supplies to plant and equipment on construction sites:

◆ 400 V three phase for supplies to major items of plant having a rating above 3.75 kW such as cranes and lifts. These supplies must be wired in armoured cables

◆ 230 V single phase for supplies to items of equipment which are robustly installed such as flood-lighting towers , small hoists and site offices. These supplies must be wired in armoured cable unless run inside the site offices

◆ 110 V single phase for supplies to all portable hand tools and all portable lighting equipment. The supply is usually provided by a reduced voltage distribution unit which incorporates splash-proof sockets fed from a centre-tapped

110 V transformer. This arrangement limits the voltage to earth to 55 V, which is recognised as safe in most locations. A 110 V distribution unit is shown in Fig. 4.25. Edison screw lamps are used for 110 V lighting supplies so that they are not interchangeable with 230 V site office lamps.

There are occasions when even a 110 V supply from a centre-tapped transformer is too high, for example, supplies to inspection lamps for use inside damp or confined places. In these circumstances a safety extra-low voltage (SELV) supply would be required.

Industrial plugs have a keyway which prevents a tool from one voltage being connected to the socket outlet of a different voltage. They are also colour coded for easy identification as follows:

440 V – red
230 V – blue
110 V – yellow
50 V – white
25 V – violet

Agricultural and Horticultural Installations

Especially adverse installation conditions are to be encountered on farms and in commercial greenhouses because of the presence of livestock, vermin, dampness, corrosive substances and mechanical damage. The 16th Edition of the IEE Wiring Regulations consider these installations very special locations and has

Figure 4.26 Farms and Commercial Greenhouses require special Consideration

devoted the whole of Section 605 to their requirements. In situations accessible to livestock the electrical equipment should be of a type which is appropriate for the external influences likely to occur and should have protection against solid objects and water splashing from any direction (Regulation 605-11-01).

Horses and cattle have a very low body resistance, which makes them susceptible to an electric shock at voltages lower than 25 V rms.

In buildings intended for livestock, all fixed wiring systems must be inaccessible to the livestock and cables liable to be attacked by vermin must be suitably protected.

PVC cables enclosed in heavy duty PVC conduit are suitable for installations in most agricultural buildings. All exposed metalwork must be provided with supplementary equipotential bonding in areas where livestock is kept (Regulation 605-08-02). In many situations, waterproof socket outlets to BS 196 must be installed. Except for SELV circuits, all socket outlet circuits must be protected by an RCD complying with the appropriate British Standard. The operating current must not exceed 30 mA and have a maximum operating time of 40 ms with a residual current of 150 mA (Regulation 605-03-01).

Cables buried on agricultural or horticultural land should be buried at a depth not less than 450 mm, or 600 mm where the ground may be cultivated and the cable must have an armour sheath and be further protected by cable tiles. Overhead cables must be installed so that they are clear of farm machinery or placed at a minimum height of 5.2 m to comply with Regulation 522-08-01 and Table 4B of the On Site Guide.

Flammable and Explosive Installations

Most flammable liquids only form an explosive mixture between certain concentration limits. Above and below this level of concentration the mix will not explode. The lowest temperature at which sufficient vapour is given off from a flammable substance to form an explosive gas-air mixture is called the *flash-point*. A liquid which is safe at normal temperatures will require special consideration if heated to

Figure 4.27 Petrol Pumps must be controlled by Flameproof Equipment because of the Potential Danger

flash-point. An area in which an explosive gas-air mixture is present is called a ***hazardous area***, as defined by BS 5345, and any electrical apparatus or equipment within a hazardous area must be classified as flameproof. Flameproof equipment is manufactured to a robust standard of construction. All access and connection points have wide machined flanges which damp the flame in its passage across the flange. Flanged surfaces are firmly bolted together with many recessed bolts, as shown in Fig. 4.28. Wiring systems within a hazardous area must be flameproof fittings using an appropriate method such as:

◆ Mineral insulated cables terminated into accessories with approved flameproof glands. These have a longer gland thread than normal MICC glands of the type shown in Fig. 4.14.

Figure 4.28 Flameproof fittings

Where the cable is laid underground it must be protected by a PVC sheath and laid at a depth of not less than 500 mm

◆ PVC armoured cables terminated into accessories with approved flameproof glands or any other wiring system which is approved by BS 5345. All certified flameproof enclosures will be marked **Ex**, indicating that they are suitable for potentially explosive situations, or **Eex**, where equipment is certified to the harmonised European Standard. All the equipment used in a flameproof installation must carry the appropriate markings, as shown in Fig. 4.29 if the integrity of the wiring system is to be maintained. Flammable and explosive installations are to be found in the petroleum and chemical industries, which are classified as group 11 industries. Mining is classified as group 1 and receives special consideration from the Mining Regulations because of the extreme hazards of working underground. Petrol filling pumps must be wired and controlled by flameproof equipment to BS 5345 and meet the requirements of the Petroleum Regulation Act 1928 and 1936 and any local licensing laws concerning the keeping and dispensing of petroleum spirit.

Figure 4.29   Flameproof equipment markings

Support and Fixing Methods for Electrical Equipment

Individual conductors may be installed in trunking or conduit and individual cables may be clipped directly to a surface or laid on a tray using the wiring system which is most appropriate for the particular installation. The installation method chosen will depend upon the contract specification, the fabric of the building and the type of installation – domestic, commercial or industrial.

It is important that the wiring systems and fixing methods are appropriate for the particular type of installation and compatible with the structural materials used in the building construction. The electrical installation must be compatible with the installed conditions, must not damage the fabric of the building or weaken load-bearing girders or joists.

Figure 4.30   Methods of Fixing Electrical Equipment

Let us look at some of the methods of fixing electrical cables and equipment.

Cable Clips

PVC insulated and sheathed wiring systems are usually fixed with PVC clips in order to comply with IEE Regulation 522-08 and Table 4A of the *On Site Guide* shown earlier in this chapter (Table 4.2). The clips are supplied in various sizes to hold the cable firmly and the fixing nail is a hardened masonry nail. Figure 4.31

shows a cable clip of this type. The use of a masonry nail means that fixings to wood, plaster, brick or stone can be made with equal ease.

Figure 4.31   PVC insulated and sheathed cable clip

When heavier cables, trunking, conduit or luminaires have to be fixed, a screw fixing is often needed. Wood screws may be screwed directly into wood but when fixing to brick, stone, plaster or concrete it is necessary to drill a hole in the masonry material, which is then plugged with a material to which the screw can be secured.

Plastic Plugs

A plastic plug is made of a hollow plastic tube split up to half its length to allow for expansion. Each size of plastic plug is colour coded to match a wood screw size.

A hole is drilled into the masonry, using a masonry drill of the same diameter and to the same length as the plastic plug (see Fig. 4.32). The plastic plug is inserted into the hole and tapped home until it is level with the surface of the masonry. Finally the fixing screw is driven into the plastic plug until it becomes tight and the fixture is secure.

Figure 4.32 Screw fixing to plastic plug

Expansion Bolts

The most well known expansion bolt is made by Rawlbolt and consists of a split iron shell held together at one end by a steel ferrule and a spring wire clip at the other end. Tightening the bolt draws up an expanding bolt inside the split iron shell, forcing the iron to expand and grip the masonry. Rawlbolts are for heavy duty masonry fixings (see Fig. 4.33).

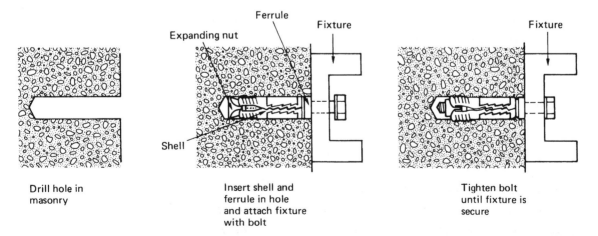

Figure 4.33 Expansion bolt fixing

A hole is drilled in the masonry to take the iron shell and ferrule. The iron shell is inserted with the spring wire clip end first so that the ferrule is at the outer surface. The bolt is passed through the fixture, located in the expanding nut and tightened until the fixing becomes secure.

For the most robust fixing to masonry material an expansion bolt, such as that made by Rawlbolt, should be used.

Spring Toggle Bolts

A spring toggle bolt provides one method of fixing to hollow partition walls which are usually faced with plasterboard and a plaster skimming. Plasterboard and plaster wall or ceiling surfaces are not strong enough to support a load fixed directly into the plasterboard, but the spring toggle spreads the load over a larger area, making the fixing suitable for light loads (see Fig. 4.34).

A hole is drilled through the plasterboard and into the cavity. The toggle wings are compressed and passed through the hole in the plasterboard and into the cavity where they spring apart and rest on the cavity side of the plasterboard. The bolt is tightened until the fixing becomes firm.

Girder Fixings

In many commercial and industrial buildings it is necessary to fix trunking, conduit and tray to the structural fabric of the building. In general, it is unacceptable to drill holes in the load-bearing structure of the

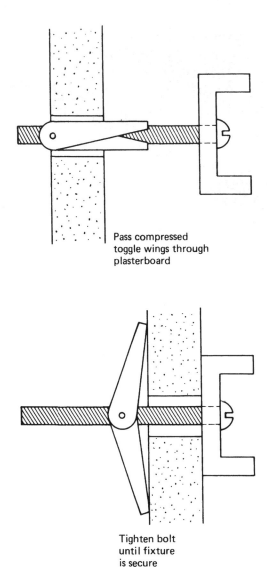

Pass compressed
toggle wings through
plasterboard

Tighten bolt
until fixture
is secure

Figure 4.34   Spring toggle bolt fixing

building to support the electrical installation for fear of weakening the building structure itself. However, spring clips or compression brackets are available which attach to the girders and hold the electrical systems securely. Figure 4.35 shows some manufactured girder supports for electrical equipment.

Figure 4.35 Girder supports

Electrical Installation, Inspection and Testing

Having fixed everything securely and completed the electrical installation, it must be inspected and tested before being put into operation. The process of inspection is a visual thing. The installation must be carefully scrutinised before being tested to ensure that it is safe to be made electrically "alive". The process of testing implies the use of instruments to obtain readings. The test results must be compared with "relevant criteria" to make sure that they are satisfactory (Regulation 713-01-01).

The tests required by the Regulations BS:7671 Requirements for Electrical Installations, must be carried out in the order given below so that safety systems are tested first. If any test indicates a failure to comply, then that test and all preceding tests must be repeated after the fault has been rectified.

1. Continuity of Protective Conductors (CPCs)

The objective of the test is to ensure that every circuit protective conductor is correctly connected and has a very low resistance.

The test is made with the supply disconnected from the consumer's earthing terminal to the farthest point of each CPC as shown in Fig. 4.36 using an ohmmeter continuity tester. The resistance of the long test lead is subtracted from the test readings to give the resistance value of the CPC.

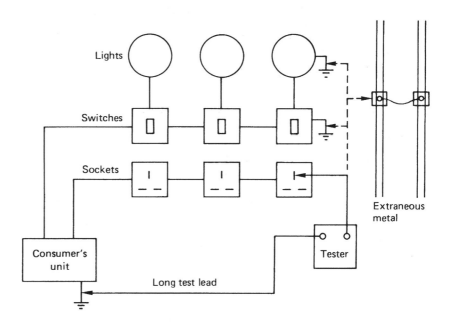

Figure 4.36

Relevant criteria tell us that a satisfactory test result would be resistance values in the order of 0.05 Ω or less.

2. Continuity of Ring Final Circuit Conductors

This test is carried out with the supply disconnected using an ohmmeter and verifies the continuity of the phase neutral and protective conductors. It also verifies that the conductors are all connected in a "ring" and that the ring has no breaks or interconnections.

3. Insulation Resistance

The object of the test is to verify the "quality" of the insulation and that the insulation resistance has a very high value. The test is made at the consumer unit with the supply disconnected using an insulation resistance meter which supplies a voltage of 500 V.

Pilot indicator lamps, discharge lighting and electronic equipment must be temporarily disconnected before this test begins to avoid false readings and possible damage to equipment as a result of the test voltage.

Relevant criteria tells us that a satisfactory test result would be a minimum resistance value of $0.5\,M\Omega$ but if values of less than $2\,M\Omega$ are recorded then this might indicate a latent but not yet visible fault in the installation which would require further investigation. A new installation would typically have an insulation resistance value of infinity (symbol ∞).

4. Polarity

The object of the test is to ensure that all fuses, MCBs and switches are connected in the phase conductor only and that all socket outlets are correctly wired.

The test is carried out with the supply disconnected using an ohmmeter as follows:

1 switch off the supply at the main switch

2 remove all lamps and unplug all equipment

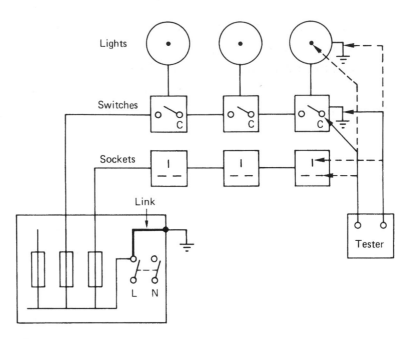

Figure 4.37

3 fix a temporary link between phase and earth on the consumer's side of the main switch as shown in Fig. 4.37.

4 test between the common terminal and earth at each switch connection

5 test between the live pin and earth at each socket outlet

6 remove the link when the test is completed.

Relevant criteria tell us that a satisfactory test result would be a very low resistance value, approaching zero ohms for each test.

When all the tests are completed and proved satisfactory, the supply may be switched on. Functional testing is then carried out to ensure the correct operation of all circuits. Functional testing means that devices are operated to confirm that they are working properly and are correctly adjusted. The integral test button marked T or Test on an RCD should be pressed to prove the mechanical parts of the RCD.

Electrical Test Instruments

Electrical installation testing in accordance with the relevant Regulations demands that we use specialist test instruments. It is unacceptable for a professional electrician to carry out electrical testing using instruments bought at the local DIY superstore. Test instruments must meet the instrument standard BSEN 61557 and carry an "in date" calibration certificate, otherwise test results are invalid.

Safe Working Environment

In Chapter 1 we looked at some of the Laws and Regulations that affect our working environment. We looked at Safety Signs and PPE and how to recognise and use different types of fire extinguishers. The structure of companies within the electrotechnical industry and the ways in which they communicate information by drawings, symbols and standard forms was also discussed.

We began to look at safe electrical isolation procedures in Chapter 1 and then discussed this topic further in Chapter 3. Safe manual handling techniques and safe procedures for working above ground level were shown in Figs 3.3 to 3.7.

This might be a good time to revise some of the Safety at Work topics discussed in Chapter 3.

◆ Read again 'Avoiding Accidents in the Workplace' on page 120
◆ Look at the 'Hazard Risk Assessment' process shown in Fig. 3.2.
◆ Read again the section on 'Safe Working above Ground Level on page 127
◆ And finally, always practice the 'Safe Electrical Isolation and Lock Off' procedure shown in Fig. 3.11.

In Chapter 3, under the heading 'Avoiding Accidents in the Workplace' we looked at the common causes of accidents at work and how to control the risks associated with various hazards. At Fig. 3.2 we looked at the 'Hazard Risk Assessment' process.

If your career in the electrotechnical industry is to be a long, happy and safe one, you must always behave responsibly and sensibly in order to maintain a safe working environment. Before starting work, make a safety assessment – what is going to be hazardous – will you require PPE – do you need any special access equipment? Carry out safe isolation procedures before beginning any work. You do not necessarily have to do these things formally, such as carrying out the risk assessment described in Chapter 3, but just get into the habit of always working safely and being aware of the potential hazards around you when you are working.

Do not leave your tools lying around for others to fall over or steal. Keep them close by you in a toolbox. The tools and equipment which you are not using should be locked away in a safe storage place.

Finally, when a job is finished, clean up and dispose of all waste material responsibly.

Correct Disposal of Waste Material

The Controlled Waste Regulations 1998 tell us that we have "a Duty of Care" to handle, recover and dispose of waste responsibly.

The Environmental Protection (Duty of Care) Regulations 1991 tell us that any business has a duty to

Figure 4.38   Disposing of waste material responsibly

ensure that any waste produced is handled safely and in accordance with the law.

Your company is responsible for the waste that it produces even after handling it over to another party such as a Skip Hire company. If such a third party mishandles your waste or disposes of it irresponsibly then it is the responsibility of the company you work for, not the Skip Hire company. The duty of care under the new Regulations has no 'time limit' and extends until the waste has either been finally and properly disposed of or fully recovered.

If you have read and understood the whole of this Chapter, you have completed all of the underpinning knowledge requirements of the Fourth Unit of the City & Guilds 2330 Syllabus for the Level 2 Certificate in Electrotechnical Technology.

When you have completed the practical assessments required by the City & Guilds Syllabus, which you are probably doing at your local College, you will be ready to tackle the on-line Assessment. So, to prepare you for the On-Line Assessment, try the following Assessment Questions.

If a material has hazardous properties, it may need to be dealt with as 'Special Waste'. Containers may be classified as 'Special Waste' if they contain residues of hazardous or dangerous substances. If the residue is 'Special' then the whole container is Special Waste.

Do not burn scrap cable on site, re-cycle it through a scrap metal merchant.

Electrotechnical companies produce very little waste material and even smaller amounts of 'Special Waste'. Most electrical contractors deal with waste by buying in the expertise and building in these costs to the total cost of a contract. However, this method still requires individuals to sort any waste responsibly by placing it in the appropriate skip or container.

To comply with the Waste Regulations:

◆ Make sure waste is transferred only to 'authorised' companies

◆ Make sure that the waste being taken is accompanied by the proper paperwork called 'waste transfer notes'

◆ Label waste skips and waste containers so that it is clear to everyone what type of waste is going into which skip or container

◆ Minimise the waste that you produce and do not leave it behind when a job is completed for someone else to clear away. As the producer of any waste, you are responsible for it. Remember there is no time limit on the Duty of Care for waste materials

Assessment Questions

Identify the statements as true or false. If only part of the statement is false, tick false

1 The Electricity at Work Regulations tell us that the IEE (called the IET from 31st March 2006) Wiring Regulations (BS 7671) is a code of practice which is widely recognised and accepted in the UK. If your electrical work meets the requirements of the IET Wiring Regulations, it will meet the requirements of all other relevant regulations.
True ❏   False ❏

2 The main reason for incorporating the Wiring Regulations into British Standards BS 7671 was to create harmonisation with European Standards.
True ❏   False ❏

3 A cable sheath earth supply or TN-S System of supply is one of the most common types of underground supply in the UK.
True ❏   False ❏

4 A **delivery note** is a standard form completed by most electrical trainees to inform an employer of how much time has been spent working on a particular job.
True ❏   False ❏

5 When materials are delivered to site, the person receiving the goods is required to sign the driver's **time sheet** to prove that the supplier has delivered the goods as requested.
True ❏   False ❏

6 A radial socket circuit is wired from the source of supply to each socket in turn and the last socket is wired back to the source of supply.
True ❑   False ❑

7 From the 1st April 2006 only the new wiring colours must be used for all fixed wiring. That is: brown for phase, blue for neutral and green combined with yellow for all single phase circuits.
True ❑   False ❑

8 Industrial installations use robust cable enclosures such as conduits and trunking. A conduit is a square or rectangular section from mild steel plate. A trunking is a tube, or pipe in which insulated conductors are contained.
True ❑   False ❑

9 Individual cables or accessories may be fixed directly to a surface with a suitable nail, screw or bolt. A spring toggle bolt provides a good method of fixing to concrete or masonry, a Rawlbolt provides a good method of fixing to hollow partition walls.
True ❑   False ❑

10 The 'Waste Regulations' tell us that we have a 'Duty of Care' to handle, recover and dispose of waste responsibly. Your company is responsible for the waste that it produces, so always make sure that waste material is put into the proper skip and taken away only by 'authorised' companies.
True ❑   False ❑

Multiple Choice Assessment Questions

Tick the correct answer. Note that more than ONE answer may be correct

11 The Electricity at Work Regulations are:
- **a** Non-statutory Regulations ❏
- **b** Statutory Regulations ❏
- **c** a code of practice ❏
- **d** a British Standard ❏

12 The IEE Regulations are:
- **a** Non-statutory Regulations ❏
- **b** Statutory Regulations ❏
- **c** a code of practice ❏
- **d** a British Standard ❏

13 A British Standard having a BS number is a:
- **a** Statutory Regulation ❏
- **b** Non-statutory Regulation ❏
- **c** British compliant Standard ❏
- **d** European harmonised Standard ❏

14 A British Standard having a BS EN number is a:
- **a** Statutory Regulation ❏
- **b** Non-statutory Regulation ❏
- **c** British compliant Standard ❏
- **d** European harmonised Standard ❏

15 Part 5 of the IEE Regulations deals with:
- **a** Protection for Safety ❏
- **b** Selection and Erection of Equipment ❏
- **c** Special Installations ❏
- **d** Inspection and Testing ❏

16 **Part 6 of the IEE Regulations deals with:**
  a Protection for Safety ❏
  b Selection and Erection of Equipment ❏
  c Special Installations ❏
  d Inspection and Testing ❏

17 **A scale drawing showing the position of equipment by graphical symbols is a description of a:**
  a block diagram ❏
  b wiring diagram ❏
  c circuit diagram ❏
  d layout diagram or site plan ❏

18 **A diagram which shows the detailed connections between individual items of equipment is a description of a:**
  a block diagram ❏
  b wiring diagram ❏
  c circuit diagram ❏
  d layout diagram or site plan ❏

19 **A diagram which shows very clearly how a circuit works, where all components are represented by a graphical symbol is a description of a:**
  a block diagram ❏
  b wiring diagram ❏
  c circuit diagram ❏
  d layout diagram or site plan ❏

20 **A Time Sheet shows:**
  a a record of goods delivered by a supplier ❏

b a record of work done which is outside
the original contract ❏

c information about work to be done,
usually a small job ❏

d the actual time spent working on
a particular job or site ❏

21 A Job Sheet or Job Card shows:

a a record of goods delivered by
a supplier ❏

b a record of work done which is
outside the original contract ❏

c information about work to be done,
usually a small job ❏

d the actual time spent working on
a particular job or site ❏

22 A Day Work Sheet shows:

a a record of goods delivered by
a supplier ❏

b a record of work done which is
outside the original contract ❏

c information about work to be done,
usually a small job ❏

d the actual time spent working on a
particular job or site ❏

23 A Delivery Note shows:

a a record of goods delivered by
a supplier ❏

b a record of work done which is
outside the original contract ❏

c information about work to be
done, usually a small job ❏

d the actual time spent working
   on a particular job or site   ❑

24 A cable sheath earth supply is also called a:
a TN-S system   ❑
b TN-C-S system   ❑
c TT system   ❑
d Standby system   ❑

25 A PME supply is also called a:
a TN-S system   ❑
b TN-C-S system   ❑
c TT system   ❑
d Standby system   ❑

26 A no earth provided supply is also called a:
a TN-S system   ❑
b TN-C-S system   ❑
c TT system   ❑
d Standby system   ❑

27 The electricity supply to a domestic consumer is usually protected at the incoming service position by a:
a Meter   ❑
b Double pole switch   ❑
c 100 A MCB   ❑
d 100 A HBC fuse   ❑

28 The assumed current demand for each lighting point in a domestic installation should be based upon the equivalent of:
a 5 amps per lampholder   ❑
b 6 amps per lampholder   ❑
c 100 Watt per lampholder   ❑
d 3 kW per lampholder   ❑

29 **The protective Type B MCB for a lighting circuit fed from a consumer unit in 1.0 mm or 1.5 mm cable should be rated at:**

   a 6 A or 10 A   ❏

   b 10 A or 16 A   ❏

   c 16 A or 32 A   ❏

   d 32 A or 40 A   ❏

30 **The protective Type B MCB for a ring circuit fed from a consumer unit in 2.5 mm cable should be rated at:**

   a 6 A or 10 A   ❏

   b 10 A or 16 A   ❏

   c 16 A or 32 A   ❏

   d 32 A only   ❏

31 **Each ring circuit of 13 A sockets must cover a floor area of no more than:**

   a 50 m^2 ❏

   b 75 m^2 ❏

   c 100 m^2 ❏

   d unlimited   ❏

32 **A radial circuit of 13 A sockets wired in 2.5 mm PVC cable must cover a floor area of no more than:**

   a 50 m^2 ❏

   b 75 m^2 ❏

   c 100 m^2 ❏

   d unlimited   ❏

33 **A radial circuit of 13 A sockets wired in 4.0 mm PVC cable must cover a floor area of no more than:**

   a 50 m^2 ❏

   b 75 m^2 ❏

c 100 m² ☐

d unlimited ☐

34 An MI cable is especially suited to:
  a domestic installations ☐
  b fire alarm installations ☐
  c burying underground ☐
  d industrial installations ☐

35 A PVC /SWA cable is especially suited to:
  a domestic installations ☐
  b fire alarm installations ☐
  c burying underground ☐
  d industrial installations ☐

36 A PVC insulated and sheathed cable is especially suited to:
  a domestic installations ☐
  b fire alarm installations ☐
  c burying underground ☐
  d industrial installations ☐

37 FP 200 cables are especially suited to:
  a domestic installations ☐
  b fire alarm installations ☐
  c burying underground ☐
  d industrial installations ☐

38 A steel conduit installation is especially suited to:
  a domestic installations ☐
  b fire alarm installations ☐
  c burying underground ☐
  d industrial installations ☐

39 **A metallic trunking installation is especially suited to:**

    a domestic installations   ❑

    b fire alarm installations   ❑

    c burying underground   ❑

    d industrial installations   ❑

40 **Cable tray installations are especially suited to:**

    a domestic installations   ❑

    b fire alarm installations   ❑

    c burying underground   ❑

    d industrial installations   ❑

41 **Bathroom installations receive special consideration in the IEE Regulations because of the hazard associated with:**

    a electricity and flammable liquids   ❑

    b electricity and water   ❑

    c the presence of corrosive substances   ❑

    d the potential for mechanical damage   ❑

42 **Construction sites receive special consideration in the IEE Regulations because of the hazard associated with:**

    a electricity and flammable liquids   ❑

    b electricity and water   ❑

    c exposure to wind and rain   ❑

    d presence of livestock and vermin   ❑

43 **Agricultural installations receive special consideration in the IEE Regulations because of the hazard associated with:**

    a electricity and water   ❑

    b presence of livestock and vermin   ❑

c　potential for mechanical damage　❏

d　electricity and flammable liquids　❏

44 Petrol pump installations receive special consideration from many Statutory Regulations because of the hazard associated with:

a　electricity and water　❏

b　electricity and flammable liquids　❏

c　exposure to wind and rain　❏

d　the temporary nature of the supply　❏

45 Locations containing a bath or shower are divided into zones or separate areas. The most dangerous zone is classified as:

a　Zone 0　❏

b　Zone 1　❏

c　Zone 2　❏

d　Zone 10　❏

46 The permissible colours of 230 V single phase fixed wiring up to 30th March 2006 was:

a　brown, blue, green and yellow　❏

b　brown, black, grey　❏

c　red, black, green and yellow　❏

d　red, yellow, green and yellow　❏

47 The new European harmonised fixed wiring colours which must be used after the 1st April 2006 for a 230 V single phase circuit are:

a　brown, blue, green and yellow　❏

b　brown, black, grey　❏

c　red, black, green and yellow　❏

d　red, yellow, green and yellow　❏

48 **PVC insulated and sheathed cables are very likely to be fixed and supported by:**
   a wood screws and plastic plugs ❏
   b a PVC clip and hardened nail ❏
   c an expansion bolt ❏
   d a clip on girder fixing ❏

49 **A lightweight piece of electrical equipment is very likely to be fixed to a plasterboard by:**
   a wood screws and plastic plugs ❏
   b a PVC clip and hardened nail ❏
   c an expansion bolt ❏
   d a spring toggle bolt ❏

50 **A heavy electric motor is very likely to be fixed to a concrete floor by:**
   a wood screws and plastic plugs ❏
   b a clip on girder fixing ❏
   c an expansion bolt ❏
   d a spring toggle bolt ❏

51 **A run of trunking suspended in an industrial installation is very likely to be fixed to the main structure of the building by:**
   a wood screws and plastic plugs ❏
   b a clip on girder fixing ❏
   c an expansion bolt ❏
   d a spring toggle bolt ❏

52 **A run of cable tray suspended in a modern Supermarket building is very likely to be attached to the main structure of the building by:**
   a wood screws and plastic plugs ❏
   b a clip on girder fixing ❏

   c an expansion bolt ❑

   d a spring toggle bolt ❑

53 **What action is necessary to produce a 'secure electrical isolation'?**

   a Isolate the supply and observe that the voltage indicator reads zero ❑

   b First connect a test device such as a voltage indicator to the supply ❑

   c Larger pieces of equipment may require isolating at a local isolator switch ❑

   d The isolated supply must be locked off or secured with a small padlock ❑

54 **A voltage proving unit:**

   a is used for transmitting data along optical fibre cables ❑

   b provides a secure computer supply ❑

   c shows a voltage indicator to be working correctly ❑

   d tests for the presence of a mains voltage supply ❑

55 **For working even a short distance above ground level for long periods, the safest piece of access equipment would be:**

   a a stepladder ❑

   b a platform tower ❑

   c an extension ladder ❑

   d a hard hat ❑

56 **An example of 'Special Waste' is:**

   a sheets of asbestos ❑

   b old fibre-glass roof insulation ❑

c old fluorescent tubes ❑

d part coils of PVC insulated cables ❑

57 Special Waste must be disposed of:

a in the general site skips ❑

b in the general site skip by someone
designated to have a 'duty of care' ❑

c at the 'Household Waste' re-cycling centre ❑

d by an 'authorised company' using a
system of waste transfer notes' ❑

**58 The Health & Safety at Work Act places the
responsibility for safety at work on:**

a the employer ❑

b the employee ❑

c both the employer and employee ❑

d the main contractor ❑

**59 Under the Health & Safety at Work Act an
Employer must ensure that:**

a the working conditions are appropriate
and safety equipment is provided ❑

b employees take reasonable care of
themselves and others as a result of
work activities ❑

c employees co-operate with an employer
and do not interfere with or mis-use
safety equipment ❑

d that plant and equipment is properly
maintained ❑

**60 Under the Health & Safety at Work Act
Employees must ensure that:**

a the working conditions are appropriate
and safety equipment is provided ❑

b they take reasonable care of themselves and others as a result of work activities ❏

c they co-operate with an employer and do not interfere with or mis-use safety equipment ❏

d plant and equipment is properly maintained ❏

Solutions to Assessment Questions

Chapter 1

1	True	**13**	c	**25**	c
2	True	**14**	a	**26**	a
3	True	**15**	b	**27**	c
4	False	**16**	d	**28**	c
5	True	**17**	a	**29**	a
6	False	**18**	b	**30**	d
7	True	**19**	c	**31**	d
8	True	**20**	c	**32**	a
9	False	**21**	d	**33**	b
10	True	**22**	a	**34**	c
11	a, b, c	**23**	c	**35**	d
12	d	**24**	c		

Chapter 2

1	True	18	c	35	a
2	False	19	d	36	a, c
3	True	20	c	37	c, d
4	False	21	c	38	c
5	False	22	d	39	b
6	True	23	c	40	b
7	True	24	a, c	41	c
8	True	25	b, d	42	d
9	False	26	c	43	d
10	False	27	b	44	d
11	c	28	b	45	c
12	b	29	b	46	a
13	d	30	a, b	47	b
14	c	31	b, c	48	c
15	b	32	b	49	a, d
16	c	33	d	50	c
17	d	34	b		

Chapter 3

1	True	21	d	41	b
2	False	22	b, d	42	a
3	False	23	b	43	d
4	True	24	c	44	b
5	True	25	a, d	45	d
6	False	26	d	46	b
7	True	27	b	47	b
8	True	28	a	48	a
9	True	29	d	49	d
10	False	30	b	50	d
11	b, c, d	31	c	51	d
12	c, d	32	a, b	52	d
13	c	33	a, c	53	b
14	a	34	a, d	54	c
15	b	35	d	55	c
16	c	36	b	56	c
17	a, b, d	37	c	57	b
18	c, d	38	d	58	c
19	c, d	39	d	59	c
20	c	40	c	60	b

Chapter 4

1 True	**21** c	**41** b
2 True	**22** b	**42** b, c
3 True	**23** a	**43** a, b, c
4 False	**24** a	**44** b
5 False	**25** b	**45** a
6 False	**26** c	**46** c
7 True	**27** d	**47** a
8 False	**28** c	**48** b
9 False	**29** a	**49** d
10 True	**30** d	**50** c
11 b	**31** c	**51** b
12 a, c, d	**32** a	**52** b
13 c	**33** b	**53** d
14 d	**34** b, c, d	**54** c
15 b	**35** c	**55** b
16 c	**36** a	**56** a, c
17 d	**37** b	**57** d
18 b	**38** d	**58** c
19 c	**39** d	**59** a, d
20 d	**40** d	**60** b, c

Index

Withdrawn